ENCYCLOPEDIA OF
BIRDS

ENCYCLOPEDIA OF
BIRDS

CONSULTANT EDITOR
Joseph Forshaw

ILLUSTRATIONS BY
Dr David Kirshner

NATURAL WORLD

Published in the United States by
Academic Press, A Division of Harcourt Brace & Company
525 B Street, Suite 1900, San Diego, CA 92101-4495, USA
Distributed worldwide exclusively by Academic Press
(except in Australia and New Zealand)

Senior Editor, Life Sciences: Charles R. Crumly Ph.D.

Conceived and produced by Weldon Owen Pty Limited
59 Victoria Street, McMahons Point, NSW 2060, Australia
A member of the Weldon Owen Group of Companies
Sydney • San Francisco

First published in 1991
Second edition 1998
Copyright © Weldon Owen Pty Limited 1998

Publisher: Sheena Coupe
Associate Publisher: Lynn Humphries
Project Editors: Jenni Bruce; Helen Cooney
Editorial Assistant: Veronica Hilton
Captions: Terence Lindsey
Index: Garry Cousins
Designers: Denese Cunningham; Sue Rawkins
Picture Research: Annette Crueger; Terence Lindsey; Grant Young
Production Manager: Caroline Webber
Production Assistant: Kylie Lawson
Vice President International Sales: Stuart Laurence

Co-ordination of scientific and editorial contributors by
Linda Gibson, Project Manager, Australian Museum Business Services

ISBN 0-12-262340-1

A catalog record for this book is available from
the Library of Congress, Washington, DC.

Printed by Kyodo Printing Co. (Singapore) Pte Ltd
Printed in Singapore

A WELDON OWEN PRODUCTION

Endpapers: Sanderlings, a species of sandpiper, foraging along the shore at sunset.
Photo by Frans Lanting/Minden Pictures
Page 1: Rainbow bee-eaters characteristically sit on conspicuous perches, waiting for
large insects such as bees, wasps, and dragonflies to fly by. Photo by Geoff Longford
Pages 2–3: Canada goose in a field of wildflowers. Photo by Philip de Renzis/The
Image Bank
Pages 4–5: A mallard, one of the most familiar of ducks, takes off from a pond.
Page 7: The roseate spoonbill is a fairly common species in the mangrove swamps of
southern USA. Photo by Barry Lipsky/Tom Stack & Associates
Pages 10–11: Mallards and their near relatives are found all over the world.
Pages 12–13: Migrating snow geese photographed at the Bosque del Apache
Wildlife Refuge, USA. Photo by Jeff Foott/Survival Anglia
Pages 44–45: Cape gannets congregate in vast flocks at their breeding grounds.
Photo by V. Serventy/Planet Earth Pictures

S. Nielsen/DRK Photo

CONSULTANT EDITOR

JOSEPH FORSHAW
Research Associate, Department of Ornithology,
Australian Museum, Sydney, Australia

CONTRIBUTORS

DR GEORGE W. ARCHIBALD
Director,
International Crane Foundation,
Wisconsin, USA

DR LUIS F. BAPTISTA
Chairman and Curator,
Department of Mammalogy and Ornithology,
California Academy of Sciences,
San Francisco, USA

DR ANTHONY H. BLEDSOE
Department of Biological Sciences,
University of Pittsburgh,
USA

DR WALTER J. BOCK
Professor of Evolutionary Biology,
Columbia University,
New York, USA

WALTER E. BOLES
Collection Manager, Bird Section,
Division of Terrestrial Vertebrates,
Australian Museum,
Sydney, Australia

PETER L. BRITTON
formerly Honorary Associate,
National Museum of Kenya, and
Editor, Checklist of the Birds of East Africa

DR P. A. CLANCEY
Research Associate,
Durban Natural Science Museum,
South Africa

DR CHARLES T. COLLINS
Professor of Biology,
California State University,
Long Beach, USA

DR JOEL CRACRAFT
Curator and Chairman,
Department of Ornithology,
American Museum of Natural History,
New York, USA

DR FRANCIS H. J. CROME
formerly Principal Research Scientist,
Division of Wildlife and Ecology,
Commonwealth Scientific and
Industrial Research Organisation,
Australia

DR JOHN P. CROXALL
Head of Bird and Mammal Section,
British Antarctic Survey,
Cambridge, England

G. R. CUNNINGHAM-VAN SOMEREN †
formerly Ornithologist Emeritus,
National Museums of Kenya, Nairobi

DR S.J.J.F. DAVIES
Adjunct Professor of the School
of Environmental Biology,
Curtin University of Technology,
Perth, Australia

DR JON FJELDSÅ
Chief Curator of Birds, Zoological Museum,
University of Copenhagen,
Denmark

DR HUGH A. FORD
Associate Professor,
Department of Zoology,
University of New England,
Armidale, Australia

CLIFFORD B. FRITH
Honorary Research Fellow,
Queensland Museum, Australia

DR STEPHEN GARNETT
Senior Principal Conservation Officer,
Queensland Department of Environment,
Cairns, Australia

DR COLIN J. O. HARRISON
formerly Principal Scientific
Officer of Ornithology,
British Museum (Natural History),
London, England

DR ALAN KEMP
Head Curator of Birds,
Transvaal Museum,
Pretoria, South Africa

DR SCOTT M. LANYON
Director, James Ford Bell
Museum of Natural History,
University of Minnesota, USA

TERENCE LINDSEY
Associate of the Australian Museum,
Sydney, Australia

DR KIM W. LOWE
Principal Policy Analyst,
Flora & Fauna Directorate,
Department of Natural Resource
and Environment,
Melbourne, Australia

S. MARCHANT
formerly Exploration Manager,
Woodside Petroleum Company,
Australia

DR H. ELLIOTT MCCLURE
formerly Director, Migratory Animal
Pathology Survey in Eastern Asia,
Walter Reed Army Institute of Research,
Washington, USA

DR PENNY OLSEN
Division of Botany and Zoology,
Australian National University,
Canberra, Australia

DR KENNETH C. PARKES
Curator Emeritus of Birds,
Carnegie Museum of Natural History,
Pittsburgh, USA

DR ROBERT B. PAYNE
Curator of Birds and Professor of Zoology,
Museum of Zoology,
University of Michigan, USA

DR MICHAEL R.W. RANDS
Director and Chief Executive,
BirdLife International,
Cambridge, England

IAN ROWLEY
formerly Senior Principal Research Scientist,
Division of Wildlife and Ecology,
Commonwealth Scientific and
Industrial Research Organisation, Australia

DR E. A. SCHREIBER
Executive Director,
Ornithological Council,
Virginia, USA

DR LESTER L. SHORT
Lamont Curator of Birds,
American Museum of Natural History, and
Adjunct Professor of Biology,
City University of New York, USA

PROFESSOR PETER J.B. SLATER
School of Environmental & Evolutionary Biology,
University of St Andrews,
Fife, Scotland

DR G. T. SMITH
Senior Research Scientist,
Division of Wildlife and Ecology,
Commonwealth Scientific and
Industrial Research Organisation,
Australia

ALISON STATTERSFIELD
Senior Research Officer,
BirdLife International,
Cambridge, England

DR FRANK S. TODD
Executive Director,
Ecocepts International,
San Diego, USA

DR EDWIN O. WILLIS
Associate Professor, Departmento de Zoologia,
Universidade Estadual Paulista,
Sao Paulo, Brazil

CONTENTS

FOREWORD

Ornithologist Oliver Austin once said, "If an animal has feathers, it's a bird. If it hasn't, it isn't." Feathers and flight. They define birds. Coupled with these, the ability of birds to sustain their internal temperature (warm-bloodedness) has enabled them, like mammals, to colonize most of the Earth.

Also like mammals, birds evolved from reptiles, but more recently—about 160 million years ago. Anatomically, all birds are remarkably alike—whether zooming hummingbird, soaring kite, flightless ostrich, or diving penguin. And all still carry a visible reptilian inheritance of scales on legs and feet. When next you view a warbler on a branch or an eagle on the wing, admire the geometric patterns of its feathers, which evolved from scales, and think reptile—just for a moment.

Although birds are tied to the land to nest, incubate their eggs, and rear their young, flight enables them to fish the farthest seas and find food amidst the cliffs and cactus of the driest deserts. Their unchallenged mastery of the skies and of the treetop realm above our heads captures our imaginations. No other living creatures can travel so far or so fast under their own power. It is no wonder that they have been a source of envy and inspiration for humans since the Stone Age.

As though flight were not enough, many birds have striking colors and forms, and voices we consider beautiful. Studies of bird song are the foundation of our understanding of animal communication. And birds have provided models for our studies of animal behavior, of territoriality and dominance, and especially of courtship. Striking plumes and capes, crests and wattles, communal dancing grounds, and dazzling behavioral displays characterize the intricate courtship of many birds; yet others gather to breed in awesome aggregations. The great colonies of flamingos, pelicans, penguins, and gannets are among the most magnificent of the world's wildlife spectacles and treasurehouses of readily gathered information about how nature works.

For human beings, acquaintance with birds opens a world of esthetic and intellectual enjoyment—but it is a world that is shrinking every day. One of every nine species of birds is now threatened with extinction because of our behavior. If there is any hope of reversing this trend, it lies in increasing public interest in and understanding of birds. That is the purpose of this book.

WILLIAM CONWAY
President and General Director
Wildlife Conservation Society, New York, USA

Angelo Gandolfi

PART ONE

THE WORLD OF BIRDS

▲ *Birds are characterized by feathers. The intricacy of these structures offers the potential for a wide range of colors and patterns among birds, from muted browns and grays to the gorgeous iridescent hue of this lesser double-collared sunbird.*

INTRODUCING BIRDS

S cientists divide the animal kingdom into several major groups for classification purposes. By far the largest group is the invertebrates: it contains about 95 percent of the millions of known species of animals, including sponges, mollusks, arthropods, and insects. Groups of vertebrates, or animals with backbones, contain the other 5 percent of known species. They can be divided roughly into fishes, amphibians, reptiles, birds, and mammals. The class Aves—birds—consists of approximately 9,000 species, grouped (in the classification system adopted by this book) into 24 orders. One order, the Passeriformes (known as passerines or songbirds), contains more than half of the known bird species. The remaining orders are known collectively as non-passerines. Birds come in all shapes and sizes, from the ostrich *Struthio camelus,* standing about 2.5 meters (8 ¼ feet) tall, to the bee hummingbird *Mellisuga helenae,* which measures less than 6 centimeters (2 ⅓ inches) from tip of bill to tip of tail and possibly is the smallest bird. There are large birds that cannot fly, small ones that can hover or fly backwards, and just about every conceivable intermediate. But it is the possession of feathers that immediately differentiates birds from other animals. All birds have feathers.

PLUMAGE

Feathers constitute the plumage of a bird. As well as providing mechanical and thermal protection, the plumage assists in streamlining the body, thereby reducing friction during flight, or when moving on the ground or through water.

There are several types of feathers, but the most important are contour feathers, which constitute the ordinary visible plumage, and down feathers, which form a hidden underlayer in most adult birds and usually constitute the plumage of newly-hatched chicks. In most species, feathers other than down grow from definite tracts of skin (the pterylae), and the intervening areas (the apteria) are bare. A bird's plumage is replaced regularly by molting.

Plumage coloration and pattern helps birds to recognize other individuals of their own species and features prominently in their displays. The dazzling array of plumage colors exhibited by birds are the result of the structural and pigmentary colors found in feathers. Structural colors are due to either interference with light (the result being iridescence) or the scattering of light, and the responsible structures are in the barbs and barbules of feathers. Pigmentary colors are widespread in birds and are due to pigments, the most common of which is melanin, the same dark pigment found in human skin, hair, and eyes. Many colors result from a combination of two or more pigmentary colors or from a combination of pigmentary and structural colors.

BILLS AND FEET

The bill and feet are other major elements of a bird's external appearance, and the enormous variation in the structure and shape of these appendages contrasts with a relative uniformity in plumage. The bill is the projecting jaws of a bird encased in horny sheaths, and it must serve the bird in the same way as our hands function for us. It is used in nest building, for preening feathers, or as a weapon in combat, but its primary function is for food-gathering. Different foods require different bill shapes.

Structural modifications are present also in the legs (tarsi) and feet. Birds that spend most of their time on the ground have long legs, well suited to walking and running, whereas tree-dwelling birds have short, stout legs to facilitate climbing and perching. Some birds, especially those occurring in cold climates, have feathered legs and even feathered toes.

No bird has more than four toes; some have three, and the ostrich is unique in having only two. In most species, three toes point forward and the first toe is turned backward, but there are exceptions, ranging from all four pointing forward, as in swifts (family Apodidae), to having two pointing forward and two turned backward, as in parrots (order Psittaciformes). There are structural

▶ Birds range in size from the tiny bee hummingbird, so small it could fit comfortably into a matchbox, to the ostrich, towering much taller than a human being; the variety of their colors and patterns is kaleidoscopic. But this variety masks a fundamental similarity in body plan, and birds vary far less in size and basic body structure than most other groups of animals.

PARTS OF A BIRD

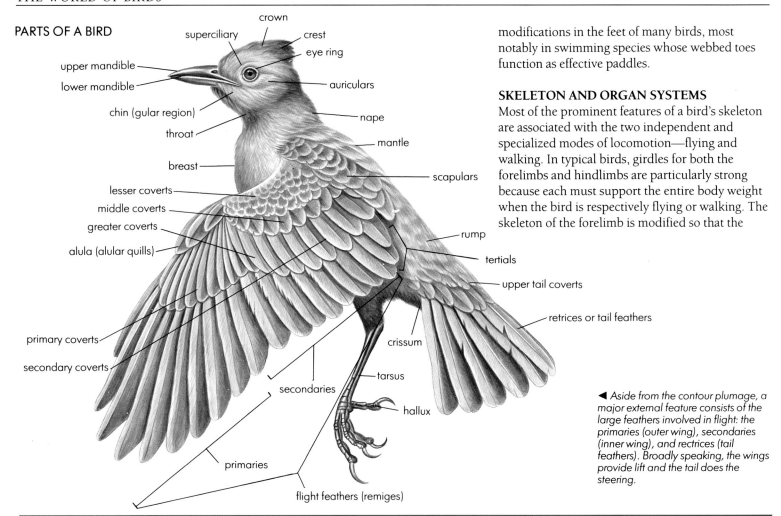

modifications in the feet of many birds, most notably in swimming species whose webbed toes function as effective paddles.

SKELETON AND ORGAN SYSTEMS

Most of the prominent features of a bird's skeleton are associated with the two independent and specialized modes of locomotion—flying and walking. In typical birds, girdles for both the forelimbs and hindlimbs are particularly strong because each must support the entire body weight when the bird is respectively flying or walking. The skeleton of the forelimb is modified so that the

◄ Aside from the contour plumage, a major external feature consists of the large feathers involved in flight: the primaries (outer wing), secondaries (inner wing), and rectrices (tail feathers). Broadly speaking, the wings provide lift and the tail does the steering.

THE FEATHERS OF A BIRD

The structure of a feather is an extremely successful and very efficient combination of lightness and strength. The feather itself is merely an elaborate, highly specialized product of the outer layer of skin, and it consists almost entirely of a very strong substance called keratin—the same substance that constitutes the hair and fingernails of humans, the fur and claws of mammals, and the scales of reptiles.

In a contour feather there are two clearly discernible components: the central spine or rachis (often referred to as the quill) and the vane or web on each side. The rachis has strong keratin walls and is filled with keratinized cells vaguely resembling the pith of some plants. Similar material fills the laterally branching barbs, which together make up the vane. Along the sides of each barb are hooked barbules, which interlock with barbules on the adjoining barb to give the feather its strength. If the interlocking barbules become unhooked, they can be re-engaged simply by stroking the feather from the base towards the tip, as is done by a bird in the act of preening.

In down feathers only the calamus or rachis base is retained. Parallel barbs project from around the top of the calamus, like spikes atop a coronet, and these give down its characteristic fluffy texture.

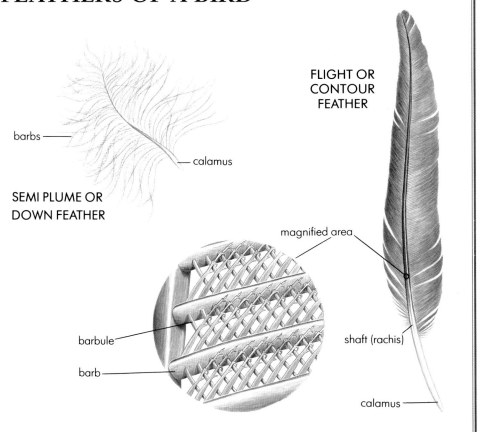

arm, forearm, and hand support the wing. The hindlimb also is specialized, in that some metatarsals (bones of the foot) are fused and lengthened so that the leg appears to contain an extra segment.

Except for the neck, the vertebral column is comparatively immobile, and many of the vertebrae are fused. The skull is light, with a compact rounded brain-case, which is filled quite tightly by the brain. The bill is attached to the skull quite flexibly, so allowing movement of the upper mandible to increase the gape; this mobility of the upper mandible is especially marked in parrots but is very slight or absent in the ratites and a few other birds. The skeleton of many birds is lightened by extensive pneumatization, a condition in which bones are hollow and contain air-sacs. Even though some bones are hollow, they are structurally very strong, having internal trusslike reinforcements which add to their strength. Lightness and strength are essential for flight.

Dominating the circulatory system is a four-chambered heart, which is different in arrangement from the four-chambered heart of mammals. Thick-walled arteries convey blood at high pressure from the heart to all parts of the body. Thin-walled veins carry the blood through a capillary network within the organs and tissues before returning it to the heart, and fluid exudates pass into the lymphatic vessels. Birds capable of sustained flight have a proportionately large heart in comparison to poor fliers or those that fly only short distances. Like mammals, birds are "warm-blooded". Although this capability of maintaining body temperature above that of the surroundings was acquired independently by the two groups, the physiology of thermoregulation for mammals and birds is remarkably similar, and is a striking example of parallel evolution.

Birds have a highly specialized respiratory system. The small lungs comprise only about two percent of the body volume, but connecting air-sacs are well developed, and in total may be up to 20 percent of the body volume. These air-sacs are located in various parts of the body, and they play an important role in the through passage of air.

The digestive tract of a bird is basically the same as that of other vertebrates and consists of a coiled tube or gut leading from the mouth to the anus. Food passes from the mouth into the gullet and then to the crop, which is a thin-walled distensible pocket of the gullet where food is stored for subsequent digestion or feeding of the young by regurgitation. The crop is well developed in grain-eating and many flesh-eating birds, less developed in other species, and absent altogether in some insect-eating birds. The proventriculus and the ventriculus or gizzard together correspond to the stomach in mammals, and again are well developed in grain-eating species. From the gizzard food passes to the duodenum and intestines, where

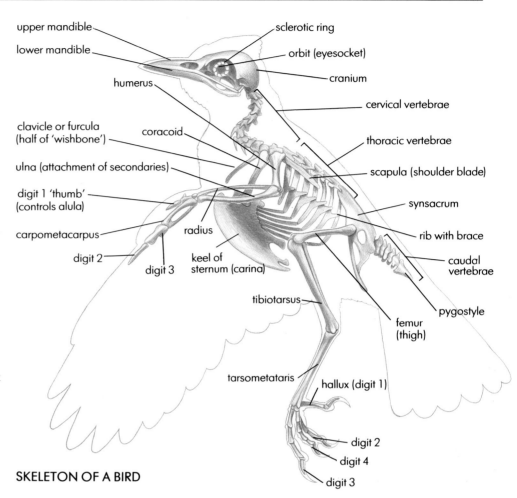

upper mandible
lower mandible
humerus
clavicle or furcula (half of 'wishbone')
coracoid
ulna (attachment of secondaries)
digit 1 'thumb' (controls alula)
carpometacarpus
digit 2
digit 3
radius
keel of sternum (carina)
tibiotarsus
tarsometataris
hallux (digit 1)
digit 2
digit 4
digit 3
sclerotic ring
orbit (eyesocket)
cranium
cervical vertebrae
thoracic vertebrae
scapula (shoulder blade)
synsacrum
rib with brace
caudal vertebrae
pygostyle
femur (thigh)

SKELETON OF A BIRD

digestion is completed before waste is excreted through the anus. Birds have no urinary bladder, so nitrogenous wastes are excreted in the form of urea, a semi-solid paste-like substance, after water has been absorbed in the cloaca. The cloaca is a common opening through which the products of the reproductive, digestive, and excretory systems are passed. Some birds, such as owls, eliminate the indigestible components of their food in the form of pellets regurgitated through the mouth.

THE SENSES

The general structure of the bird eye is similar to that found in all vertebrates. However, the extremely well-developed, efficient eyes possessed by almost all birds, especially the large eyes of some birds of prey and nocturnal species, lead ornithologists to conclude that vision is of the utmost importance.

Attempts to ascertain the level of hearing possessed by birds have met with only partial success, but the few auditory functions that have been measured are almost as sensitive as they are in humans. A higher proficiency was detected in the ability to recognize different sounds repeated so rapidly that to the human ear they become inextricably fused. Birds apparently possess adequate olfactory organs, but in some species the sense of smell seems to be poorly developed and plays little part in their lives.

▲ The skeleton of a bird, although broadly similar to that of other vertebrates, is highly modified to support powered flight. The bones of the hand, for example, are fused to form what is essentially a single digit, which supports the main flight feathers. Other typically avian features include the backward-pointing tabs ("uncinate processes") on each rib, and a prominent keel-like structure on the sternum or breastbone, which serves to anchor the enormous pectoral muscles supplying power to the wings.

Cross section of bone

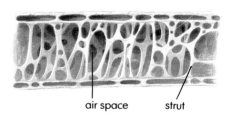

air space   strut

▲ The long bones of the limbs (and in some birds many other bones in the body) are thin-walled and hollow, but intricately braced and strutted inside to provide the maximum strength and rigidity with the least premium in weight.

A brood of baby lesser black-backed gulls hatches at a colony in Wales:

Kathie Atkinson

▶ *A baby bird breaks free of the eggshell with the help of the "egg-tooth" on the tip of the bill. Obvious only at hatching, this feature will be shed or re-absorbed within a day or so.*

Kathie Atkinson

▶ *Damp and bedraggled, the newly-hatched chick rests briefly after its exertions. Chicks of some species utter piping calls while still within the egg, and it has been shown that these calls hasten the hatching of other chicks in the brood, improving the odds that all will emerge together.*

Kathie Atkinson

▶ *Birds like this, whose chicks hatch down-covered, open-eyed, and mobile, are known as precocial. Some can even feed themselves immediately. Even so, it may be several days before the baby bird is fully capable of regulating its own body temperature, and it will require frequent brooding by its parents to shield it from rain and extremes of hot and cold weather.*

HOW BIRDS REPRODUCE

All birds lay eggs, within which development of the embryo subsequently takes place, but of course this form of reproduction is prevalent in other groups of animals. As in the majority of vertebrates, the adult male has testes and the female has ovaries, although in nearly all bird species, only the ovary on the left side is functional. During copulation the cloaca of both sexes is everted so that sperm can transfer from male to female, but in some birds (for example, many ducks and the ratites) part of the cloaca of the male is modified to form a penis. Fertilization of released ova takes place in the upper oviduct, then as each egg passes along the oviduct, layers of albumen are deposited on it. In the wider and greatly distensible uterus, the shell and pigment are added to complete the egg, which finally passes through the vagina and cloaca to be expelled into the nest.

The egg must be kept at the correct temperature for embryonic development. This is usually brought about through contact with the body of a parent, and the adults of many species develop brood-patches—areas denuded of feathers and richly supplied with blood vessels. The parent bird settles on the nest so that its brood patch or patches cover the egg or eggs; and with regular changeovers of the parents or short breaks away for feeding, the eggs are incubated until the chicks hatch. Incubation periods vary, from 80 days for a large albatross, to 10 days for some small passerines. Some species do not incubate their own eggs but parasitize other birds by laying their eggs in the host's nest, and the megapodes make use of natural sources of heat such as sunlight or the fermentation of decaying vegetation to maintain the temperature of eggs buried in a mound or sand.

JOSEPH FORSHAW

John Daniels/Ardea London Ltd

▶ *This brood of Eurasian robins is almost ready to fledge (that is, become fully feathered and capable of flight). The Eurasian robin is an altricial bird: the chicks at hatching are naked, blind, and helpless, and dependent on their parents for food and care for several weeks before they can leave the nest.*

CLASSIFYING BIRDS

The diversity of a group of animals such as birds could not be covered easily in this book without the use of a well-established classification. It enables biologists to summarize a vast amount of biological information in an efficient fashion. For example, categorizing birds as members of the animal kingdom, the phylum Chordata, the subphylum Vertebrata, and the class Aves informs us that each individual bird possesses, among other things, gill slits, a dorsal hollow nerve cord, a vertebral column, a neural crest, feathers, and forelimbs modified into wings.

AVIAN CLASSIFICATION

The class Aves is divided into orders—24 in the classification adopted for this book, although there are additional orders containing only extinct birds. Knowing that a bird is an owl, order Strigiformes, informs us that it possesses a bony arch on its radius bone, among other features. And if a bird is a hornbill, family Bucerotidae, we know that the first two cervical vertebrae are fused into a single unit. Each order is subdivided into families (about 165), each family into genera (slightly over 2,000), and each genus into species (slightly over 9,000).

It must be emphasized that considerable disagreement still exists about the limits of some groups and the relationships between orders and between families included in an order. However, most of the family-level groups covered in this book are well substantiated and will retain their identity even when we learn more about the relationships of birds to each other.

WHAT IS A SPECIES?

A species is defined as a group of actually or potentially interbreeding populations of organisms, which are genetically isolated from other such groups. Species maintain their separation from each other by the possession of intrinsic isolating barriers which prevent the exchange of genetic material among them.

The scientific name given to a species is made up of two words derived from Greek or Latin—for example, *Falco peregrinus*. This name is used by ornithologists around the world, no matter whether they are Dutch taxonomists working in Egypt, Russian researchers in Siberia, or Spanish-speakers in South America. However, the vernacular or common name that people give to this species can vary from place to place; thus the peregrine falcon may also be called the black-cheeked falcon in English, and different names in many other languages, but they all refer to one species, *Falco peregrinus*.

Where populations of a species are separated geographically they may develop slightly different details of size or plumage color, and can be identified as separate races or subspecies. A third word is then added to the scientific name. For example, *Falco peregrinus tundrius* is one of the North American subspecies, and *Falco peregrinus calidus* one of the European subspecies. An adult female should be able to breed successfully with an adult male of a related subspecies where their geographic ranges overlap and the habitat conditions are suitable.

From time to time biologists have to decide whether geographic representatives should be considered as subspecies of a species or as separate species. Such decisions are largely arbitrary, as there are no objective tests for judging the specific status of geographic representatives whose ranges do not meet.

THE BASIS OF CLASSIFICATION

Biological classification, the work of systematists, has two main goals. The first is recognition of the basic units of biological diversity—species and their subunits (mainly subspecies); this then establishes the extent of diversity throughout all organisms, living and extinct. The second is the arrangement of these basic units into a system of increasingly higher-level groups, providing the foundation for summaries of biological knowledge. When two or more species are quite similar in their morphology, physiology, behavior, and ecology, they can be classified in the same genus; all species in a genus are presumed to be descendants of a common ancestor.

Biologists are interested in a single natural classification suitable for all comparative analyses, one which reflects the past evolutionary history of organisms by summarizing the amount of evolutionary change along each lineage and the splitting of lineages. These aspects would be revealed by the number of taxa (groups) at successive levels—such as species, genus, family, order, class—indicating the degree of relationships among species and higher-level taxa.

Several different types of biological classification have been used in recent years. One type is known as phenetic. The only aspect of evolutionary

Morten Strange

▲ Birds which superficially resemble each other are not necessarily related, while differences in outward appearance often mask a common lineage. Thus this rhinoceros hornbill shares many features with kingfishers despite its different appearance, and both are grouped in the order Coraciiformes.

correlated with each other, so evolutionary taxonomists must decide which aspect should be given greater importance in particular cases. The order Passeriformes, for example, includes more than half of the known species of living birds, for its members show much less diversity than the rest of the orders of birds combined. In contrast, the order Coliiformes (mousebirds) contains a single genus with only six species but it is quite distinct from other orders. Evolutionary classifications contain the greatest amount of information for biologists, and provide the best all-purpose general reference system. The evolutionary classification used in this book follows closely the system advocated in the 16 volumes of the recently completed Peters' *Check-list of Birds of the World.*

ESTABLISHMENT OF A CLASSIFICATION
The establishment of a biological classification is a two-step process. First is the formulation of hypotheses about the classification of groups—for example, that the kingfishers (family Alcedinidae) and the hornbills (family Bucerotidae) are members of the same group, the order Coraciiformes. This hypothesis is tested scientifically against taxonomic properties of characters, of which the most important is homology. The second step is the formulation of hypotheses about the taxonomic properties of characters—homology, for example—which are tested against empirical observations. This second step, character analysis, is the most important part of classifying organisms into higher-level groups.

The words homology, homologous and homologue come from the Greek *homologos* meaning "agreeing, corresponding". In biological usage a homologue is a feature in two or more organisms that stems phylogenetically from the same feature in the immediate common ancestor of these organisms. Thus the hypothesis that the fused first and second cervical vertebrae are homologous in species of hornbills means that this feature was inherited from such fused vertebrae in the immediate ancestor of all known hornbills. Hypotheses about homologous features are tested by comparing them and ascertaining their similarities. These similarities are assumed to be paternal ones—descriptive of the feature in the immediate common ancestor and remaining unchanged during the evolution to each descendent organism. Thus, homology of the fused cervical vertebrae in hornbills would be tested by establishing similarities in the structure of this feature in diverse hornbill species. This is the only available valid way to test hypotheses about homologies.

Unfortunately this test is frequently not very robust and often does not provide correct answers. Hence further analyses are needed to establish a degree of confidence in each homologue. This involves functional and adaptational analyses of the

change reflected in such a classification is the amount of evolutionary change expressed as the similarity among taxa. A second type is cladistic (from the Greek *klados* meaning "branch"). The only evolutionary aspect reflected in this type of classification is the splitting of phyletic lineages (branching patterns). A third type, evolutionary classification, attempts to summarize both the amount of evolutionary change in phyletic lineages and the splitting of these lineages; however, these two evolutionary aspects are not absolutely postulated homologues, and estimates of the probability that two similar features evolved independently ("convergent evolution").

The possibility of being fooled by independent evolution of unrelated organisms subjected to similar demands from their environment can be reduced by studying various features so that at least some of the features will be independent of similar selective demands. For this reason, the scientist will attempt to use a diversity of features, choosing them carefully to include those associated with different aspects of the life of the organisms. So although both grebes and loons have webbed feet, the presence of different types of webbing suggests that these two groups are not closely related despite being foot-propelled diving birds. Systematists are more confident in the correctness of a classificatory hypothesis if it is supported by a variety of homologous features. But each feature must be carefully and independently analysed.

Ornithologists have used this approach, but with varying success. The major problem appears to be a great emphasis placed on finding new taxonomic characters—biochemical ones during the past two decades—but in the absence of functional/adaptational analyses to establish how much confidence should be given to taxonomic characters in different groups of birds.

Moreover, the tendency has been for each systematist to emphasize the classification supported by the characters he or she used. After all, most of the classic morphological characters used to establish the currently accepted classification have never been properly analysed. And neither have the newly established biochemical and genetic (DNA) characters.

WHAT IS A SEQUENCE?
What is a classification and what is a sequence? Why does the sequence of birds vary in different books? Before addressing these questions, we should consider the difference between a classification and a sequence.

Classifications are systems expressing the evolutionary relationships of taxonomic groups arranged in an inclusive, non-overlapping hierarchy. In any one taxon, all members descend from a single common ancestor. The taxa in this type of hierarchy are arranged in a series of categories at different levels; for birds, the class Aves is the highest categorical level, followed by orders, families, genera, and species. Intermediate levels such as superfamilies, subfamilies, and tribes are also used.

Sequences are arrangements of the taxa to suit books and other data banks with similar linear restrictions. Rules do exist for the establishment of sequences—such as more primitive groups being listed before more advanced groups—but other equally valid sequences could be established from the same classification. Broadly accepted standard sequences are important because they permit greater ease of communication. For this book we have adopted not only the basic classification but also the general sequence used in Peters' *Check-list* because it is the most standard recent sequence for birds of the world.

RULES FOR SCIENTIFIC NAMES
The International Code of Zoological Nomenclature is concerned with names for groups at different levels, from subspecies to families, with the goal of establishing a stable universal set of taxonomic names for all animals.

Priority means that the valid name of any taxon is the oldest name applied to it. If new studies reveal that two species are members of the same genus but were formerly classified in separate genera, they should both be given the generic name that was published first. However, priority is only one of the rules used to achieve stability and universality in zoological nomenclature. Long-term established usage regulated through plenary powers of the International Commission on Zoological Nomenclature is another.

Special care has been used in this volume to use the valid name for each avian taxon, especially those advocated in the recently developed list of names for bird families.

WALTER J. BOCK

Leo Meier/Weldon Trannies

▲ *Most Coraciiformes, such as this white-throated kingfisher, nest in cavities and have large bills and colorful plumage. One structural feature that suggests their common lineage is the unusual condition known as syndactyly, in which the three forward-facing toes are fused together along part of their length.*

ORDERS AND FAMILIES OF BIRDS
The following list is based on *Check-list of Birds of the World* by J.L. Peters, E. Mayr, J.C. Greenway Jr., *et. al.*, 1931–1987, 16 volumes, Cambridge, Massachusetts, Museum of Comparative Zoology.

CLASS AVES

ORDER STRUTHIONIFORMES	**RATITES AND TINAMOUS**	**ORDER PROCELLARIIFORMES**	**ALBATROSSES AND PETRELS**	**ORDER SPHENISCIFORMES**	**PENGUINS**
Struthionidae	Ostrich			**Spheniscidae**	Penguins
Tinamidae	Tinamous				
Rheidae	Rheas	**Diomedeidae**	Albatrosses	**ORDER**	
Casuariidae	Cassowaries	**Procellariidae**	Shearwaters	**GAVIIFORMES**	**DIVERS**
Dromaiidae	Emu	**Hydrobatidae**	Storm petrels	**Gaviidae**	Divers (loons)
Apterygidae	Kiwis	**Pelecanoididae**	Diving petrels		

ORDER
PODICIPEDIFORMES **GREBES**
Podicipedidae Grebes

ORDER
PELECANIFORMES **PELICANS & THEIR ALLIES**
Phaethontidae Tropicbirds
Pelecanidae Pelicans
Phalacrocoracidae Cormorants, anhingas
Sulidae Gannets, boobies
Fregatidae Frigatebirds

ORDER
CICONIIFORMES **HERONS & THEIR ALLIES**
Ardeidae Herons, bitterns
Scopidae Hammerhead
Ciconiidae Storks
Balaenicipitidae Whale-headed stork
Threskiornithidae Ibises, spoonbills
Cathartidae New World vultures

ORDER
PHOENICOPTERIFORMES **FLAMINGOS**
Phoenicopteridae Flamingos

ORDER
FALCONIFORMES **RAPTORS**
Accipitridae Osprey, kites, hawks, eagles, Old World vultures, harriers, buzzards, harpies, & buteonines
Sagittariidae Secretarybird
Falconidae Falcons, falconets, caracaras

ORDER
ANSERIFORMES **WATERFOWL & SCREAMERS**
Anatidae Geese, swans, ducks
Anhimidae Screamers

ORDER
GALLIFORMES **GAMEBIRDS**
Megapodiidae Megapodes (mound-builders)
Cracidae Chachalacas, guans, curassows
Phasianidae Turkeys, grouse, etc

ORDER
OPISTHOCOMIFORMES **HOATZIN**
Opisthocomidae Hoatzin

ORDER
GRUIFORMES **CRANES & THEIR ALLIES**
Mesitornithidae Mesites
Turnicidae Hemipode-quails (button quails)
Pedionomidae Collared hemipode (plains wanderer)
Gruidae Cranes
Aramidae Limpkins
Psophiidae Trumpeters
Rallidae Rails
Heliornithidae Finfoots
Rhynochetidae Kagus
Eurypygidae Sunbittern
Cariamidae Seriemas
Otididae Bustards

ORDER
CHARADRIIFORMES **WADERS & SHOREBIRDS**
Jacanidae Jacanas
Rostratulidae Painted snipe
Dromadidae Crab plover
Haematopodidae Oystercatchers
Ibidorhynchidae Ibisbill
Recurvirostridae Stilts, avocets
Burhinidae Stone curlews (thick knees)
Glareolidae Coursers, pratincoles
Charadriidae Plovers, dotterels
Scolopacidae Curlews, sandpipers, snipes
Thinocoridae Seedsnipes
Chionididae Sheathbills

Laridae Gulls, terns, skimmers
Stercorariidae Skuas, jaegers
Alcidae Auks

ORDER
COLUMBIFORMES **PIGEONS & SANDGROUSE**
Pteroclididae Sandgrouse
Columbidae Pigeons, doves

ORDER
PSITTACIFORMES **PARROTS**
Cacatuidae Cockatoos
Psittacidae Parrots

ORDER
CUCULIFORMES **TURACOS & CUCKOOS**
Musophagidae Turacos, louries (plaintain-eaters)
Cuculidae Cuckoos, etc

ORDER
STRIGIFORMES **OWLS**
Tytonidae Barn owls, bay owls
Strigidae Hawk owls (true owls)

ORDER
CAPRIMULGIFORMES **FROGMOUTHS & NIGHTJARS**
Steatornithidae Oilbird
Podargidae Frogmouths
Nyctibiidae Potoos
Aegothelidae Owlet nightjars
Caprimulgidae Nightjars

ORDER
APODIFORMES **SWIFTS & HUMMINGBIRDS**
Apodidae Swifts
Hemiprocnidae Crested swifts
Trochilidae Hummingbirds

ORDER
COLIIFORMES **MOUSEBIRDS**
Coliidae Mousebirds

ORDER
TROGONIFORMES **TROGONS**
Trogonidae Trogons

ORDER
CORACIIFORMES **KINGFISHERS & THEIR ALLIES**
Alcedinidae Kingfishers
Todidae Todies
Momotidae Motmots
Meropidae Bee-eaters
Coraciidae Rollers
Upupidae Hoopoe
Phoeniculidae Wood-hoopoes
Bucerotidae Hornbills

ORDER
PICIFORMES **WOODPECKERS & BARBETS**
Galbulidae Jacamars
Bucconidae Puffbirds
Capitonidae Barbets
Ramphastidae Toucans
Indicatoridae Honeyguides
Picidae Woodpeckers

ORDER
PASSERIFORMES

Suborder Eurylaimi **BROADBILLS & PITTAS**
Eurylaimidae Broadbills
Philepittidae Sunbirds, asitys
Pittidae Pittas
Acanthisittidae New Zealand wrens

Suborder Furnarii **OVENBIRDS & THEIR ALLIES**
Dendrocolaptidae Woodcreepers
Furnariidae Ovenbirds
Formicariidae Antbirds
Rhinocryptidae Tapaculos

Suborder Tyranni **TYRANT FLYCATCHERS & THEIR ALLIES**
Tyrannidae Tyrant flycatchers

Pipridae Manakins
Cotingidae Cotingas
Oxyruncidae Sharpbills
Phytotomidae Plantcutters

Suborder Oscines **SONGBIRDS**
Menuridae Lyrebirds
Atrichornithidae Scrub-birds
Alaudidae Larks
Motacillidae Wagtails, pipits
Hirundinidae Swallows, martins
Campephagidae Cuckoo-shrikes, etc
Pycnonotidae Bulbuls
Irenidae Leafbirds, ioras, bluebirds
Laniidae Shrikes
Vangidae Vangas
Bombycillidae Waxwings
Hypocoliidae Hypocolius
Ptilogonatidae Silky flycatchers
Dulidae Palmchat
Prunellidae Accentors, hedge-sparrows
Mimidae Mockingbirds, etc
Cinclidae Dippers
Turdidae Thrushes
Timaliidae Babblers, etc
Troglodytidae Wrens
Sylviidae Old World warblers
Muscicapidae Old World flycatchers
Maluridae Fairy-wrens, etc
Acanthizidae Australian warblers, etc
Ephthianuradae Australian chats
Orthonychidae Logrunners, etc
Rhipiduridae Fantails
Monarchidae Monarch flycatchers
Petroicidae Australasian robins
Pachycephalidae Whistlers, etc
Aegithalidae Long-tailed tits
Remizidae Penduline tits
Paridae True tits, chickadees, titmice
Sittidae Nuthatches, sitellas, wallcreeper
Certhiidae Holarctic treecreepers
Rhabdornithidae Philippine treecreepers
Climacteridae Australasian treecreepers
Dicaeidae Flowerpeckers, pardalotes
Nectariniidae Sunbirds
Zosteropidae White-eyes
Meliphagidae Honeyeaters
Vireonidae Vireos
Emberizidae Buntings, tanagers
Parulidae New World wood warblers
Icteridae Icterids (American blackbirds)
Fringillidae Finches
Drepanididae Hawaiian honeycreepers
Estrildidae Estrildid finches
Ploceidae Weavers
Passeridae Old World sparrows
Sturnidae Starlings, mynahs
Oriolidae Orioles, figbirds
Dicruridae Drongos
Callaeidae New Zealand wattlebirds
Grallinidae Magpie-larks
Artamidae Wood swallows
Cracticidae Bell magpies
Ptilonorhynchidae Bowerbirds
Paradisaeidae Birds of paradise
Corvidae Crows, jays

BIRDS THROUGH THE AGES

T he biology of living birds is better known than for any other group of vertebrates. Our understanding of the evolutionary history of birds is not as far advanced, however, in part because fewer fossils have been found. But many spectacular fossil finds in recent years, along with comparative studies of the anatomy and genetic structure of species alive today, are providing valuable evidence with which to reconstruct the pattern of avian evolution, especially the early history of birds.

"FEATHERED DINOSAURS"

Although birds arose more than 150 million years ago, the first 50 million years of avian history has yielded relatively few fossils. Those that do exist have provided important information about the modernization of the avian body plan, especially as it pertains to the evolution of flight.

During the past decade, paleontologists have reassessed the relationships between birds and two-legged theropod dinosaurs, a group that includes perhaps the most famous dinosaur of them all, *Tyrannosaurus rex*. The realization that birds are "feathered dinosaurs" arose as a result of the discovery of new specimens of the oldest known bird, *Archaeopteryx lithographica*. Indeed,

the two beautiful, nearly complete specimens of *Archaeopteryx* now housed in museums in London and Berlin were among the very first fossil birds to be described.

All six specimens of *Archaeopteryx* were found in the Solnhofen limestones of Bavaria, in southern Germany, and had lived during the late Jurassic period, 200 to 145 million years ago. The close relationship of *Archaeopteryx* and other birds to theropod dinosaurs is significant, because it means that the most distinctive characteristic of birds—flight, and all the structural modifications necessary for flight—arose in ancestors that were swift two-legged predatory creatures designed to exploit terrestrial environments rather than trees.

Although *Archaeopteryx* is very clearly related to birds, much of its skeleton recalls that of a theropod. Most of the bones of the forelimb, for example, had none of the special modifications developed in modern birds, such as the fusion of the bones in the wrist and hand facilitating the attachment and fine manipulation of the flight feathers. The bones of the shoulder girdle—the scapula and coracoid—were like those of theropods, and the two clavicles were fused into a U-shaped furcula, or "wishbone". It was long thought that the furcula was a distinctive avian feature, bracing the two shoulder girdles during flight, but we now know that it was present in

◄ *About the size of a crow or a pigeon, the earliest known bird,* Archaeopteryx lithographica, *lived in Europe over 145 million years ago.*

some advanced theropods, where it also probably functioned as a brace, but for arms used in prey capture not flight.

From the days of its discovery in 1861, *Archaeopteryx* was considered a bird because its body was covered with feathers arranged in feather tracts similar to those of modern birds. Yet despite the presence of feathers, paleontologists have debated whether *Archaeopteryx* was capable of powered flight or was simply a glider. The detailed shapes of the flight feathers suggest the former, for they were asymmetric—the leading edge was very much narrower than the trailing edge. Such a configuration appears to be correlated with a well-developed aerodynamic function and thus suggests that *Archaeopteryx* was capable of strong flight. In addition, the brain of *Archaeopteryx,* especially the portions associated with motor activity and coordination, was enlarged and also indicates a major refinement in locomotor behavior.

THE CRETACEOUS PERIOD

Some of the oldest forms of the Cretaceous period (145 to 65 million years ago) are only a little bit younger than *Archaeopteryx* itself. Despite this, all Cretaceous birds are much more like modern

▶ A number of birds inhabited the seas of the late Cretaceous period some 70 million years ago, among them several species of Hesperornis, a fish-eating bird that may have been the ancestor of modern grebes. This bird had lost the ability to fly and, like Archaeopteryx but unlike all modern birds, it had teeth.

birds, and less like theropods, and most of them were undoubtedly capable of strong flight. At the same time, their anatomy provides evidence of their dinosaur ancestry. Among the most primitive were two species whose fossils have recently been discovered near Las Hoyas, Spain. Both were the size of small songbirds and had forelimbs and shoulder girdles that were relatively similar to modern birds, whereas the skeletal elements of the pelvis and hindlimbs were more similar to *Archaeopteryx* and theropods. These birds also had a pygostyle—a fusion of the most posterior caudal vertebrae, providing a central point of attachment for the tail feathers—and an ossified sternum, which indicate improvements in the ability to fly.

Several other early Cretaceous birds are interesting because they suggest that anatomically more advanced forms had already evolved by this time. One of these, *Ambiortus,* found in Mongolia, was the size of a small crow. *Ambiortus* possessed a well-developed keel on the sternum and a very modern shoulder girdle, and (unlike the Las Hoyas fossils) the bones of the wrist were fused to form a single element, the carpometacarpus, just as in modern birds. Although nothing is known about the pelvis and hindlimbs of *Ambiortus,* another early Cretaceous bird fills in some of the gaps in our knowledge. This form, named *Gansus* after the Gansu Province of China where it was discovered, is known only from the distal part of the leg, but those bones are of an entirely modern aspect and thus are much more advanced than the Las Hoyas birds. *Gansus* was apparently a small shorebird-like species.

Perhaps the most famous of all Cretaceous fossil birds were *Hesperornis* and *Ichthyornis* from the late Cretaceous of North America. Hesperornithiform birds are also known from the early Cretaceous of England and the late Cretaceous of South America, which suggests they were successful and widespread. *Hesperornis* and *Ichthyornis* are notable because they retained the primitive theropod condition of having teeth on their upper and lower jaws, as did *Archaeopteryx.* Although many other Cretaceous birds probably had teeth, no evidence for this has yet been found. *Hesperornis* and its allies were flightless, foot-propelled divers. As such they had lost the keel on the sternum, greatly reduced the size of wing, and developed non-pneumatic bones. *Ichthyornis,* in contrast, was a strong flier, reminiscent in some respects of modern gulls, but much more primitive and unrelated to them.

THE RISE OF MODERN BIRDS

Modern birds, called the Neornithes ("new birds"), are divisible into two well-defined groups. The first of these is the paleognaths ("ancient jawed", in reference to their somewhat primitive skull) including the tinamous of South and Central America, and large flightless ratite birds such as the ostrich of Africa, the rheas of South America, the emu and cassowaries of Australia–New Guinea, and the kiwis of New Zealand. Paleognaths have had a long history that predates the breakup of Gondwana and the drift of the southern continents. Several fossils from the late Cretaceous period have been found in Mongolia and Europe. In the Paleocene (65 to 57 million years ago) of Europe and North America there existed numerous species of relatively small paleognaths, most of which were capable of powered flight.

The second group of modern birds is the neognaths ("new jawed", in reference to their more advanced skull anatomy). More than 99 percent of all species alive today are neognaths. We know that most of the major groups were represented in the Eocene (57 to 37 million years ago) and Oligocene (37 to 24 million years ago), but few of them have a fossil record from as early as the Cretaceous, primarily because of the scarcity of sediments containing fossils. Paleontologists are probably justified in inferring that many of these groups, or their ancestors, extended well into the Cretaceous.

One of the more primitive lineages of neognaths includes the galliform birds (chickens, pheasants, quails) and the anseriform birds (ducks, geese, swans). Both are first known from fossils in the Eocene, and each apparently had a worldwide distribution by that time. This radiation also included a small group of very large birds, some more than 2 meters (6½ feet) tall: the flightless diatrymas of the North American, European, and Asian Eocene. Although it was assumed for a long time that they were fierce predators, recent studies suggest they were herbivores, which would be consistent with their apparent relationship to anseriforms. It is also possible that another group of flightless giants, the dromornithids or "Mihirung birds" of Australia, are members of this lineage.

Recent evidence suggests that waterbirds such as penguins, loons, grebes, pelicans, cormorants and their allies, and the albatrosses, shearwaters and their allies comprise a distinct evolutionary lineage. Many of these groups have fossil representatives in the Eocene, and so it is reasonable to assume this radiation had its beginnings in the Cretaceous. Penguins are well represented in the fossil record of Australia, South America, and New Zealand, where they live today, and even in the Eocene they were already specialized for "flying" through water.

Few lineages of birds have as interesting a fossil record as the pelecaniforms. The most bizarre group was the pseudodontorns ("false-toothed birds"), a diverse assemblage of albatross-like gliders. All had jaws with teeth-like bony projections, which were presumably used for capturing prey while skimming the ocean's surface. Some pseudodontorns were truly gigantic, with a wingspan of as much as 6 meters (20 feet), far larger than any living albatross.

The most fascinating order of birds from a paleontological perspective is the Gruiformes, which includes rails and cranes, as well as a number of morphologically distinct families. The gruiforms have perhaps the best fossil record of any order of birds. One lineage now represented only by the trumpeters and seriemas of South America was much more diverse; it included several closely related families that radiated extensively in Europe and North America during the Eocene and Oligocene. Another lineage radiated in South America as the spectacular phorusrhacoid birds. These included a large number of gigantic species, most, if not all, of which were flightless and roamed the savannas and pampas as fierce predators. They survived to the end of the Pliocene (5 to 2 million years ago). Inexplicably, one of the largest members of the group, *Titanis walleri,* has also been discovered from Pliocene fossil records in Florida, the only

▲ New Zealand was the home of the moas, a group of large to very large wingless birds that probably existed from the Pleistocene to within 200 to 300 years ago. About a dozen species are known; Dinornis maximus, portrayed here, stood an estimated 3 meters (10 feet) tall.

▲ Fossils of the Pleistocene period (2 million to 10,000 years ago) include many species that are alive today, but many others are now extinct. The large, vulture-like Teratornis merriami lived in western North America, and many specimens have been found at the Rancho La Brea tar pits in California.

and Asia as well as in North and South America. But unquestionably the most spectacular vulture-like birds—their relationships are also obscure—were the teratorns ("wonder birds") of the Miocene (24 to 5 million years ago) and Pliocene (5 to 2 million years ago) of South America and the Pleistocene (2 million to 10,000 years ago) of North America. *Teratornis merriami* found in the Rancho La Brea tar pits in California was quite large, having a wingspan of perhaps 3.8 meters (12½ feet). But soaring across the Miocene skies of the South American pampas was *Argentavis magnificens,* the largest-known flying bird. *Argentavis* had a wingspan that may have reached 7.5 meters (24½ feet) and a body weight of 80 kilograms (176 pounds). Present evidence based on skull morphology suggests this giant was largely a predator rather than a scavenger like most vultures.

The second group of raptors, the owls, was remarkably diverse: no fewer than three families, now extinct, are known from the Paleocene and Eocene of North America and Europe, and numerous extinct genera of barn owls are known from the same deposits. Owls in the modern family Strigidae are first known from the early Oligocene, and since then have radiated nearly worldwide.

Other orders and families of birds, while not having as extensive paleontological records as the preceding groups, are represented by fossils which indicate they too had their origins more than 50 million years ago. Included among these are nightjars and their allies (Caprimulgiformes), cuckoos (Cuculiformes), parrots (Psittaciformes), swifts (Apodiformes), and kingfishers, rollers, and their allies (Coraciiformes).

The largest order of birds is the Passeriformes, or songbirds. More than 70 families and thousands of species are alive today, but because virtually all of them are small tree-dwelling birds, their fossil record is relatively poor; the fossils that have been found are only tens of thousands of years old at most. Although a small number of species have been described from deposits that are at least 40 million years old, their relationships are not clear. But genetic distances among their most divergent families are much greater than among many non-passerine groups that were present 50-60 million years ago. This suggests that songbirds arose long before that time.

All the evidence suggests that most orders and many families of living birds probably originated in the Cretaceous, more than 65 million years ago, and then subsequently radiated. It may be difficult to document this conclusion directly until we find more fossils in non-marine sediments of the late Cretaceous. Nevertheless, recent years have seen a significant growth in the field of avian paleontology, which is certain to expand our knowledge of avian evolution in the very near future.

JOEL CRACRAFT

certain record of these birds north of Brazil.

Because they inhabit aquatic environments, charadriiform birds—shorebirds, gulls, terns, and their allies—are well represented in the fossil record. At least four extinct families of the late Cretaceous of North America are tentatively placed in this order, thus attesting to the ancient origins of this group. Many contemporary families, including sandpipers, plovers, avocets, puffins, and auks, were present by the Eocene.

Another lineage of aquatic forms includes flamingos, storks, and ibises, and they too have a relatively good fossil record. All three were present and widely distributed by 45-50 million years ago.

The two great groups of raptorial birds include the falconiforms (hawks, eagles, falcons, and vultures) and the owls. Whether they are all closely related has been the subject of intense debate. Although the fossil record does not help solve this problem, it documents that both groups have been in existence for at least 50 million years. Hawks and eagles had diversified on most continents by the Eocene, and other falconiform groups such as ospreys and secretarybirds are nearly as old. The "New World" cathartid vultures, whose evolutionary relationships are still uncertain, also have an extensive fossil record, including a number of forms in the Eocene and Oligocene of Europe

HABITATS & ADAPTATIONS

The habitat of a bird can be loosely defined as the environment it occupies, particularly the climate and vegetation. Its habitat must provide food, foraging sites, cover from predators and the weather, and nesting sites. Birds have adapted to habitats as diverse as the Arctic tundra, the Sahara Desert, the Amazonian rainforest, and the middle of the oceans. They have carved out niches from the available resources. Most have survived ice ages and periods of great aridity, causing the expansion and contraction of their favored habitats. It is true that some became extinct, but others evolved to take their place. Each habitat has a characteristic array of species, many of which will display morphological adaptations to that particular habitat.

HABITAT REQUIREMENTS

A bird's habitat may be restricted by geographical barriers; for example, numerous families of songbirds such as cotingas, manakins, antbirds, and woodcreepers are restricted to South and Central America; bowerbirds, fairy-wrens, and lyrebirds are found only in the Australia-New Guinea region. More often, habitat restriction comes about because a species requires a particular resource to be present. This may be food, such as the seeds of spruce trees for the common crossbill *Loxia curvirostra* in Scandinavia, or protea flowers for the Cape sugarbird *Promerops cafer* in South Africa. Often it is a safe nesting site, such as a hole in a living pine tree for the red-cockaded woodpecker *Picoides borealis* in southeastern United States, and termite mounds for the golden-shouldered parrot *Psephotus chrysopterygius* in northern Australia.

Birds may be generalized or specialized in their habitat. The peregrine falcon *Falco peregrinus* and the barn owl *Tyto alba* occupy a wide range of habitats around the world. Kirtland's warbler *Dendroica kirtlandii* is an example of a highly specialized species, living only in jack pine woodlands recovering from fire that burnt through them six to thirteen years previously. Some species occupy different habitats in different parts of their range. The horned or shore lark *Eremophila alpestris* breeds in the high Arctic, the cold deserts of Central Asia, and the mountains of the Balkans, Morocco, and the Middle East. In North America the species is widespread in tundra, mountains, and deserts, as well as fields and grasslands. There is even a small isolated population in the Andes.

In many regions the presence of other animals, particularly predators, parasites, and competitors, may deter birds from an otherwise suitable habitat. Introduced predators such as stoats and rats have eliminated many native birds from the main islands of New Zealand, so the stitchbird *Notiomystis cincta* and the saddleback *Philesturnus carunculatus,* for example, are now restricted to

tiny offshore islands. An introduced mosquito which carries avian malaria has forced the endemic honeycreepers of the Hawaiian Islands to retreat to the mountain forests of each island. Evidence for competitive exclusion is more difficult to find, but many species expand their habitat in the absence of a similar species. For example, the horned lark is probably so widespread in North America because there are no other true larks there.

▼ *The golden-shouldered parrot is a rare bird that inhabits dry savanna woodlands on Cape York peninsula, Australia. It feeds almost entirely on seeds on the ground, and builds its nest only in termite mounds. Much sought after by aviculturists, its future is threatened by illegal trapping.*

G. Longford

Tom & Pam Gardner

▲ *Pardalotes are largely restricted to the eucalypt forests and woodlands of Australia, foraging for insects in the canopy foliage. This is the most widespread species, the striated pardalote. It nests in tree-cavities or in tunnels on the ground.*

▶ *(Opposite page) The seas are vast but suitable nesting islands for seabirds are small and scattered; many species congregate to nest in huge, crowded colonies, like these king penguins on Macquarie Island.*

Many migratory birds occupy quite different habitats in the breeding and non-breeding seasons. Seabirds roam across the oceans but breed on islands and cliffs. Perhaps the most remarkable is the marbled murrelet *Brachyramphus marmoratus* which nests in the crowns of forest trees 50 meters (150 feet) high, inland from the Pacific coast of North America.

ECOLOGICAL NICHES

The ecological niche is the role an animal occupies in its habitat—its relation to food, shelter, and enemies. Generally it is the feeding behavior that tends to determine a species' niche. For example, the great spotted woodpecker *Picoides major* of Eurasia is a forest-dwelling, bark-gleaning and probing insect- and seed-eater; and the eastern meadowlark *Sturnella magna* of North America is a grassland-inhabiting, ground-gleaning insect-eater. Of course these are little more than sketches of the birds' lives. We could add that the woodpecker requires rotten trunks to excavate its nest sites, it sometimes takes nestling birds, and occasionally feeds on the ground. But the ecological niche is usually used as a quick way of pigeonholing a

species' position in its community.

Niches are perhaps best seen when comparing the species living in any one habitat. In a forest, for example, there are seed-eaters, fruit-eaters, numerous insect-eaters, sometimes nectar-eaters, and consumers of vertebrate prey. Some species combine different foods, such as fruit and invertebrates eaten by thrushes in North America and Europe, or insects and nectar by honeyeaters in Australia. Different species taking the same type of food often forage on different levels within a habitat, known as microhabitats; for example, among insect-eaters there are different species feeding on the ground, among foliage, on bark, and capturing prey in the air. Species may take similar foods in the same place but of different sizes.

Scientists studying the ways in which birds share or partition the resources in similar communities around the world have noticed two things. First, there are often similar-looking (but possibly unrelated) birds filling similar niches on different continents; this is known as convergence. Second, there are differing numbers of species present in any given habitat.

SPECIES DIVERSITY

A birdwatcher in spring in a North American, European, or southern Australian forest might expect to see 40 or 50 bird species on a good morning. More could be added in subsequent visits, but an observer would be very pleased if their bit of forest attracted as many as 100 species during the year. However, a small area of rainforest in New Guinea may be the home of more than 200 bird species, and the best parts of Amazonia may support 300 to 500 species. Why are tropical rainforests so rich? It seems that there are several answers. Rainforests hold a greater variety of resources, such as a vast array of fruits and flowers and large insects, and foraging opportunities in vine tangles and palm fronds. Birds such as parrots, fruit-doves, jacamars, motmots, and oropendolas exploit these niches. In South America there is a whole range of species that follows army ants, capturing the insects they displace.

Islands typically have fewer kinds of birds than continents. Generally the smaller the island and the more distant it is from the mainland the fewer species it possesses. The reason that larger islands have more species is because they have more habitats and hence more ecological niches. Also each species that finds a suitable niche can become common enough to be reasonably safe from extinction. Remote oceanic islands may have few species partly because not many birds have reached them. However, recent research on fossil remains on the Pacific islands has revealed that they had many more species when they were settled by Polynesians than when Europeans arrived. So their impoverished wildlife today is partly because the island peoples exterminated many local forms.

sharp-billed or vampire finch — blood

tree finch — insects

warbler finch — insects

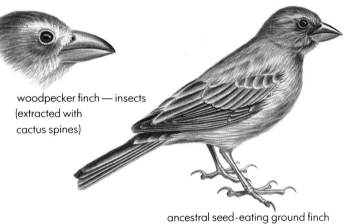

woodpecker finch — insects
(extracted with
cactus spines)

ancestral seed-eating ground finch

large cactus ground
finch — cacti

large ground finch — seeds

tree finch — plant material

▲ Arising from a single ancestral species that probably looked very much like the bird in the center, Darwin's finches have evolved into a number of species that now occupy most islands and most habitats in the Galapagos Islands.

▶ Just over half of the total body length of the sword-billed hummingbird is made up of the extraordinary bill. An inhabitant of the high Andes in South America, this hummingbird gathers nectar from a number of plant species characterized by very deep trumpet-shaped flowers.

dislodge prey from inaccessible positions. A further species, the warbler finch *Certhidea olivacea,* adopted the way of life of a warbler and gleans small insects from foliage and twigs. Perhaps most remarkable of all, one of the finches, the ground or sharp-billed finch *Geospiza difficilis,* now feeds on blood from seabirds. It perches on nesting boobies, pecks at the base of their wing and tail feathers, and laps up the blood that oozes out.

The differences in appearance of the finches on this arid, remote archipelago led Charles Darwin to the belief that species were not immutable, but can change over time—in other words, evolve. The process by which a single species diversifies into a whole array of forms exploiting different niches is known as adaptive radiation. It can be seen on many groups of oceanic islands. The Hawaiian islands provide perhaps the supreme example. Again, a finch arrived some millions of years ago, perhaps from Asia, and found an environment with endless opportunities. The result was about 40 species of honeycreepers or sicklebills (family Drepanididae), which exploited all sorts of seeds, insects, and nectar over all the main island. Sadly, the familiar story of extinction under the impact of Polynesian and European settlement followed; direct predation, habitat clearing and burning, and introduced predators and parasites have severely depleted the number of honeycreepers.

Just as adaptive radiation can occur on islands, so too it can occur on continents. Mountain ranges and deserts may act as barriers that isolate populations, allowing them to diverge.

ADAPTIVE RADIATION

Island birds include some remarkable forms. These have evolved as a result of long periods of isolation from mainland relatives in a strange environment, but also because many other mainland forms are absent. Frequently one kind of bird has diversified in ecology and appearance, so that there are now many, and these occupy different ecological niches.

This can be seen best in Darwin's finches of the Galapagos Islands. There were probably no other land-birds when the first finches arrived from the neighboring South American mainland. But there were insects, fruits, flowers and buds, as well as the seeds that make up the main diet of finches. Some of the finches continued to occupy seed-eating niches, so the larger islands have large-, medium-, and small-billed species (*Geospiza magnirostris, G. fortis* and *G. fuliginosa*), which eat large, medium, and small seeds respectively. One species evolved a longer bill and feeds on cactus flowers and fruits as well as seeds (*G. scandens*). Another evolutionary line led to the tree-dwelling finches (genus *Camarhynchus*), which feed on fruit, buds, seeds, and insects. The woodpecker finch exploits insects from beneath bark and in rotten wood; it does this by using a small stick or cactus spine to locate and

John S. Dunning/Ardea London Ltd

BILLS, FEET, WINGS, AND TAILS

Birds have characteristic sizes and shapes of their body parts, which have been adapted to suit their feeding behavior and the niches they occupy. Birds' bills display adaptation best. The long dagger-shaped bill of the herons, the huge pouches on the pelican's bill, the hooked bill of predatory birds, and the deep, heavy bills of seed-eating parrots and finches are all adaptations to their food. Differences between bills of related species reflect differences in ecology. The blue tit *Parus caeruleus* in Europe has a deeper bill than the coal tit *P. ater* as it takes insects from broad-leaved trees, while the latter feeds in conifers; a fine bill offers an advantage when probing into pine needles. A similar pattern is shown in North America, with the plain titmouse *P. inornatus* having a deeper bill for deciduous trees and the chestnut-backed chickadee *P. rufescens* having a finer bill for coniferous trees.

Feet, too, show adaptations to a bird's feeding behavior and environment. The talons of raptors for gripping large prey, the webbed feet of ducks for swimming, and the extremely long toes of the jacana for walking on waterlilies are good examples. Ground-feeding birds usually have long legs and toes, whereas tree-creeping birds have very long toes and claws but short legs. Aerial birds such as swallows and swifts have small feet; indeed the scientific name for one group of swifts is *Apus,* meaning "no feet". Legs may also be short in birds in very cold climates, to reduce heat loss; the tundra-dwelling ptarmigans (genus *Lagopus*) have very short, feathered legs and feet.

Mark Newton/Auscape International

Wings and tails can be important too. Long wings provide economy during flight. The house martin *Delichon urbica* uses far less energy when flying than similar-sized but more terrestrial birds. Short wings give maneuverability, however, and are found in birds living in dense habitats or those that indulge in aggressive aerobatics. A comparison of two species, of which one is a migrant and the other is not, shows that the former invariably has longer wings. Long tails can also provide maneuverability and are shown by most flycatching birds. They can also be important in display, as in pheasants.

Of course, many birds change their diet and even their habitat during the year, so their bills, legs, and wings have to be compromises. Many birds eat seeds and insects, but the former require a deep bill and the latter a fine bill. It seems that birds are adapted morphologically to the time when food is in shortest supply. The chaffinch *Fringilla coelebs* and the great tit *Parus major* feed on insects in summer but nuts and seeds in winter. They have fairly deep bills, adapted to when food is scarce. The long bills of many waders are poorly adapted for insects, which they eat in the breeding season, but are ideal for probing estuarine mud in winter.

Natural selection operates on birds' bills, legs, and wings by favoring those with the most efficient size and shape. These birds will survive best and leave the most offspring, who will have inherited their parents' advantageous characteristics. Natural selection takes place in this way over thousands and millions of years, and this evolution allows species to adapt to changing conditions.

HUGH A. FORD

▲ Birds that seek food by wading in shallow water generally have long legs, long necks, and long bills, like this black-winged stilt. Stilts are found in shallow wetlands of all kinds in temperate and tropical regions throughout the world.

BIRD BEHAVIOR

The behavior of birds is governed primarily by their senses of vision and hearing. In this respect they are very like humans, which probably goes part of the way—along with their beautiful plumage and striking songs—toward explaining why birds are so attractive to us. Although the behavior of birds is wonderfully varied, all of them must find their way from one place to another, find food, avoid being eaten by predators, breed with a mate of their own kind, and rear young which are well equipped to achieve all these feats in their turn. The senses play an important role in all of these activities.

EYES AND EARS LIKE HUMANS

The eyesight of most birds is rather like our own, although recent evidence suggests that some of them see very much better in ultraviolet light than we do. Likewise, their hearing range is similar to ours, but some, such as owls, have special abilities that are remarkable. The barn owl can home in on and kill a mouse in a pitch-black room within seconds because its ears are adapted for extremely accurate sound location. Most birds also resemble us in having a poor sense of smell, but again there are some exceptions: the New Zealand kiwis are noted for their ability to smell out prey.

The other main factor to be considered as a background to discussing bird behavior is their movement patterns. Birds are extremely mobile. The power of flight enables them to travel long distances in pursuit of food or mates. Those that breed in higher latitudes need not hibernate or eke out a precarious existence during the short and cold winter days, when many foods are absent or in short supply. They can travel to more equable climates where living is easier.

FINDING FOOD

The ability to find a reasonably constant supply of food is obviously very important to a flying animal that must be light and therefore cannot store large reserves. Whether they are diving for fish, probing at the water's edge for crabs, gleaning insects from the forest canopy, or searching for seeds in the undergrowth, most birds spend a high proportion of their waking hours in pursuit of food.

Not many birds cooperate in the search for food. They may be solitary hunters, like hawks or owls, or they may gather in groups where food is in abundant supply, as do finches or penguins, but they do not often assist each other to catch it.

▼ *A river kingfisher emerges triumphantly from the water with its catch. This colorful species lives along wooded rivers and streams, intently scanning the water for prey from a series of favorite low perches along the banks.*

Australian Picture Library/ZEFA

Pelicans swimming in formation, and cormorants diving in synchrony, probably help to round up fish shoals, but close cooperation to track down and kill a single prey, like that of wolves or lions, is rare among birds. The social hunting displayed by Harris's hawks, where several birds combine to catch large prey such as a rabbit, may be an example. Members of a mated pair may forage together, and there is evidence that some birds that roost together at night may benefit by gaining information from each other about where best to feed. But, when feeding, most birds look after themselves alone.

Small birds, which do not have extensive food reserves, must feed very actively through most of the daylight hours. Indeed, some very small ones, such as hummingbirds, have so little in reserve that they lose heat overnight and rely on the warmth of the sun to get them going again in the morning.

It is of great benefit to birds to find food as quickly and economically as possible, especially if they feed in the open where their searching may expose them to the danger of being eaten themselves. A good deal of evidence suggests that birds do indeed feed in this way. A thrush that has just found a worm will search in the same area more carefully—a good strategy, given that worms usually occur in groups. A flycatcher, which eats small insects in the treetops early in the day, moves closer to the ground later on when large flies become active, as these yield more energy for the work expended in catching them.

Individual birds may also develop different feeding skills and concentrate on the foods to which they are best adapted. Some gulls may feed on the shore, eating crabs and other invertebrates, while others search for food on agricultural land, and yet others specialize in the spoils to be found

▲ Talons extended, a barn owl is captured in its final approach for a landing. A cosmopolitan species, this owl has remarkably keen hearing: laboratory experiments have shown that it is quite capable of catching a mouse in utter darkness, guided solely by the tiny sounds made by its prey in breathing and moving about.

on rubbish dumps. Even with a single type of food, techniques may differ. It is not easy to prize apart the two shells of a mussel and so gain the meat inside, but oystercatchers have various different ways of doing it. Some specialize in "stabbing", inserting their bill between the valves and cutting the muscle that holds them together. The favored technique for others is "hammering", by which they break their way in through the shell. Some gulls and crows have yet another method: they fly up into the air and drop the mussel repeatedly until it breaks. They may even choose hard surfaces on which to do this, so that it is more likely to be successful.

Dropping shells onto a hard surface is just one stage removed from the use of tools. Song thrushes use special stones, their "anvils", on which they repeatedly smash snail shells until they break. Egyptian vultures take the opposite approach to break the very thick-shelled eggs of ostriches, gaining access by hitting them with a heavy stone thrown from their bill. Even more subtle is the behavior of the woodpecker finch from the Galapagos Islands. It uses a cactus spine held in the bill to extract grubs from holes in trees.

Some birds have overcome the lack of a constant food supply by storing it. A marsh tit that finds a rich supply of seeds will hide them one at a time in the surrounding area, remembering their locations and returning to eat them one at a time during the next couple of days. Other birds, such as the acorn woodpecker, use food storage as a longer-term strategy to tide them over the months when the nuts that they eat are scarce.

AVOIDING BEING EATEN

As well as feeding, birds must avoid being fed upon. A lone finch foraging in the open is easy prey to a cat or hawk. It is probably largely for this reason that many small birds feed in flocks where they can benefit from the warning provided by more pairs of eyes. In a large group one of them is bound to have its head up, looking around for danger. The first to spot a predator often produces an alarm call and so warns the others. Each bird may be able to spend more time feeding and less looking out for predators simply because of the safety in numbers. An ostrich, for example, feeding on its own, will raise its head and look around more often than when it is in a group.

Solitary birds have several ways of avoiding being preyed upon. The snipe, a secretive wading bird, sits tight until one is almost upon it, and then darts into the sky with a zig-zag flight that would be very hard for a predator to follow. Its plumage, streaked in various shades of brown, matches the long grass of the marshes where it lives, and like many cryptically colored birds its main defense comes from the difficulty predators have in detecting it. The burrowing owl in North America has an even better trick. It lives down the burrows of ground squirrels, and if one of these should chance upon it, it has a call that closely resembles that of a rattlesnake. The squirrel does not stay around to find out who produced the call!

COURTING AND MATING

The breeding behavior of birds is wonderfully diverse. Most birds are monogamous, and the male often defends a territory in which sufficient food may be found for the pair and their young. In songbirds the male may sing to attract a mate and to keep rivals out of his territory; he will threaten male intruders and court female ones with displays that often show off brightly colored parts of his plumage. But this general picture hides a wealth of variety. For example, in some species, such as phalaropes, which are small waders nesting high in the Northern Hemisphere, it is the female that courts the male. In some species that are colonial, such as gannets, the territory is only large enough to contain the nest, and feeding is done elsewhere. In some species males form "leks"—groups of very small territories on which they display to attract females to mate, then the females themselves nest elsewhere. Some males may mate with several females on the same territory, some may have several territories with a different female on each. In some songbirds, such as the house sparrow, the

▼ Many birds use courtship displays of various kinds to cement their pair bonds. These complex rituals are often beautiful or spectacular. Here a pair of western grebes perform their serene and graceful "weed dance".

Gary Nuechterlein

Brian Chudleigh

male has no real song, whereas in others, such as the superb lyrebird, the nightingale, or the brown thrasher, he may have hundreds or even thousands of phrases. These lavish songs, like the tail of a peacock, are thought to have evolved because females find the variety attractive, and the bigger and better the display the more mating success the male achieves.

After pairing comes nest-building, mating, and egg-laying. The nest may be an elaborate affair, such as those made by weavers, whose construction involves a complex sequence of movements and carefully chosen materials, or it may be a simple scrape. Some birds build no nest at all, whereas others create a vast pile. The mound of vegetation, several meters across, amassed by the male brush turkey is the most notable, especially as the eggs are incubated by the warmth of its fermentation rather than by either of the parents. In more conventional cases incubation may be by the male, by the female, or by both. Either or both of the sexes may carry out care of the young as well, with the added variant that many species have now been found to have helpers at the nest: additional birds, often the offspring from earlier broods, which help the parents to feed their chicks

and so to raise more young. Helping like this is perhaps at its most bizarre in the white-rumped swiftlet, where the female lays two eggs several weeks apart. By the time the second is laid the first has hatched and is old enough to incubate it.

Birds with young or eggs have a particular problem in keeping predators away from their brood. Special alarm calls produced when a hawk is spotted may serve to warn the mate and young so that they keep quiet and still until the danger has passed. Some small birds on the nest hiss and rattle their wing feathers in frightening similarity to a nasty predator such as a snake. Wading birds often produce distraction displays. When a dangerous intruder such as a human comes near the nest, they run away with wings dragged along the ground as if badly wounded. When the predator has been tempted to chase them far from the nest, they rise in the air and fly strongly away. The predator is distracted from one meal by the prospect of another, and as a result loses both.

Predators may also be driven off. The fulmar, a seabird from the North Atlantic which eats foods such as squid and jellyfish, as well as offal thrown from trawlers, defends its nest by spitting its last oily and half-digested meal at an intruder. It can

▲ An Australasian gannet returns to the colony with nesting material. Most birds build nests in which to cradle and protect their young; some nests are extraordinarily intricate structures taking days or even weeks to complete, while others may be little more than a rough heap of litter on the ground.

Belinda Wright

▲ *Solicitous and protracted care of their young is a very conspicuous feature of bird behavior. Here a painted stork brings food to its nestlings in India.*

▼ *The downy young of western grebes spend much of their early lives snugly riding on the back of one or other of their parents. In grebes both sexes co-operate in parental care, but male involvement is by no means universal among birds. In many species the female rears her brood unaided.*

Gary Neuchterlein

and must be fed in the nest until they have developed sufficiently to fledge. This can often be several weeks. In other birds (for example, ducks and gulls) the young are well developed at birth, can stand within a short period, and leave the nest within a day of hatching. They rapidly learn to feed themselves, and the parents provide them with only shelter and protection.

Experience plays an important role in the development of all birds, particularly after they leave the nest. A young gull learns the call of its own parent and will perk up to beg only when it hears the call of that particular adult returning to the colony. It also learns to peck at its parent's bill to obtain food. In chickens and ducks, where the young follow the mother in search of food, they learn her features also. At first a young chick will follow any large object it sees, but after a few days it has imprinted on its mother and will follow only her. Other large objects are now frightening to it and if one should appear, instead of following it the young bird will run to its mother for safety.

By experience with their parents and their siblings young birds also learn the more general features of their own species, and, when they are more mature, it is for these that they look when seeking a mate. This is true of songbirds too, even though they remain in the nest for longer. If a young male zebra finch is reared by a pair of Bengalese finches, it will prefer to court and mate with a Bengalese when it becomes adult.

As they grow, young birds develop many of the skills required to survive and reproduce by trial and error and by interactions with their parents and others around them. Young mammals, particularly predatory ones such as cats, often romp around in rough-and-tumble play. Play is less obvious in birds, but young predators such as falcons will often fly at each other and grapple in dazzling aerobatic displays. One theory is that such predators, in their play, are learning the complex skills required to capture and subdue their prey.

BIRD BRAINS

Much of the learning shown by birds is more a case of special abilities matched to a way of life than a sign of wide-ranging intelligence. A brown thrasher will master more than a thousand song phrases; a marsh tit can memorize the locations of hundreds of hidden seeds. But neither could manage the other task, nor many other tasks that to us seem simple. Natural selection has endowed them with special abilities where they need them.

This is not to argue that birds are stupid. They have large brains like those of mammals, and their behavior is elaborate and varied. The evidence for their learning by imitation is as good as that for any mammal besides ourselves and our closest relatives. This is just one of the many striking features of bird behavior that make it a fascinating subject.

PETER J. B. SLATER

score a hit at up to 2 meters (6½ feet), and its aim is deadly, so intruders are best to keep away. They are also well advised to keep clear of the territories of skuas. These gull-like seabirds dive-bomb large animals that stray near their nests, swooping down from above to approach at high speed. Though they seldom hit, only those with a steely nerve remain close by.

GROWING UP

In some birds (for example, songbirds and birds of prey) the young hatch small, naked, and helpless,

ENDANGERED SPECIES

More than a thousand bird species are threatened with extinction today. This is the startling conclusion from research by BirdLife International (formerly the International Council for Bird Preservation). In fact, the situation is worse than even this figure implies, for many more are declining or are potentially vulnerable and could soon be threatened with extinction too.

THE GLOBAL THREAT

All types of birds are at risk—passerines, non-passerines, big birds, small birds, landbirds, seabirds—but one family with many threatened species, more than 80 species in total, is the parrot family, Psittacidae. Beautiful birds like the world's largest parrot, the hyacinth macaw *Anodorhynchus hyacinthinus* from central Brazil, eastern Bolivia,

CONSERVATION WATCH

This book lists endangered species throughout Part Two, Kinds of Birds (from page 46). The first page of each chapter features a colored Key Facts panel, which includes the Conservation Watch heading.

The conservation information in the Key Facts panels is based on the *1996 IUCN Red List of Threatened Animals*, a co-publication of the IUCN World Conservation Union and Conservation International.

The level of threat is indicated by the symbols below.

!!! Critically endangered
!! Endangered
! Vulnerable
■ Other information

Keith & Liz Laidler/Ardea London Ltd

◀ *Worldwide habitat destruction and poaching have reduced many parrots to critically low populations. This is especially true of the large macaws, like this hyacinth macaw—largest of all parrots—of South America. Large birds generally need large territories, and the combination of spacious habitat requirements, low reproductive rates and long generation times, ease of capture, high cash value, and inadequate protection make it increasingly difficult for these birds to survive in the wild.*

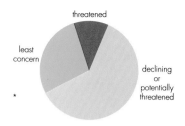

Status of world's birds

threatened

least
concern

declining
or
potentially
threatened

*

▲ The precarious status of the world's birds: about 11 percent of the world's total avifauna must be regarded as threatened.

Countries with most threatened birds

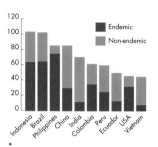

■ Endemic
■ Non-endemic

Indonesia, Brazil, Philippines, China, India, Colombia, Peru, Ecuador, USA, Vietnam

▲ The political connection: many of the world's poorest nations also have the highest total of threatened bird species.

Birds extinct since 1600
Total: 105

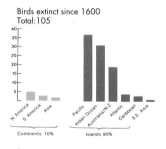

N. America, S. America, Asia, Pacific, Indian Ocean, Australia/N.Z., Atlantic, Caribbean, S.E. Asia

Continents 10%   Islands 90%

*

▲ The geography of extinction: the grim catalog to date suggests that birds of small oceanic islands are most at risk, closely followed by tropical rainforest species.

* BirdLife International

▶ One of the largest and most impressive of all birds of prey, the harpy eagle is threatened by widespread destruction of the forests of Central and South America. It is now rare and considered potentially threatened.

and northeastern Paraguay, or the world's only flightless parrot, the kakapo *Strigops habroptilus,* now confined to a couple of islands off New Zealand, could disappear if efforts are not made to save them.

Threatened birds come from all walks of avian life—some live in wetlands, others in deserts, many inhabit tiny islands—but the habitat that is home to the highest number of threatened species is tropical rainforest. Many of the threatened parrots, pigeons, pheasants, birds of prey, and hornbills are forest-dwelling, and some families such as the antbirds (Formicariidae) of Central and South America, the broadbills (Eurylaimidae) of Africa and Asia, and the birds of paradise (Paradisaeidae) of Indonesia, New Guinea, and Australia, are restricted almost entirely to forests.

Birds are threatened in all corners of the globe, on all continents, and on most islands. But three countries stand out from the others: Indonesia, Brazil, and the Philippines together account for nearly 25 percent of all threatened bird species.

Since 1600 at least 100 species, mostly island birds, have died out. More than 30 have disappeared this century. Of course extinction is a natural process: species have always died out, and others have diversified in response to a changing environment. As a consequence of evolution, the birds of today bear little resemblance to their *Archaeopteryx*-like ancestors which flew on earth some 140 million years ago. The big difference between the species that have become extinct in historical times, and those that disappeared in more ancient eras, is that recent extinctions can be attributed almost wholly to human beings.

This extinction crisis is not confined to birds. Their demise signifies that other wildlife in the same ecosystems—mammals, reptiles, invertebrates—are likely to be in trouble too. We are in fact facing the possibility of a massive loss of global biodiversity within the next few generations, unless we act now to conserve the natural world.

WHY BIRDS ARE ENDANGERED

Causes of extinction and current threats vary, depending on the type of bird, the habitat it lives in, and the region it inhabits. The greatest cause of bird extinction has been the effect of introduced species on island birds. The greatest overall threat to birds today, however, is undoubtedly habitat destruction, which affects both island species and continental species.

Habitat destruction

Almost all major habitat types have been affected by encroachment: grasslands have been widely plowed or subjected to intensive overgrazing by livestock; wetlands have been drained and converted to farmland; tropical forests have been degraded, chopped down, and burnt. It is estimated that tropical forests have already been reduced by more than 40 percent of their original area, and the

destruction continues apace. Until recently the most authoritative estimate of deforestation in the tropics was 11.4 million hectares (44,000 square miles) per year, but it is now thought that deforestation is much greater than previously estimated in Brazil, Costa Rica, India, Myanmar (Burma), the Philippines, and Vietnam. Forest clearing has also increased sharply in Cameroon, Indonesia, and Thailand. If these new studies are accurate, the world is losing up to 20.4 million hectares (78,700 square miles) of tropical forest each year.

In some areas there has been almost complete deforestation. For example, because of deforestation the island of Cebu in the Philippines has lost 39 bird species—half of the forest-dependent endemic birds once resident on the island. In other places the forest has become fragmented. Some species can survive partial clearance of their habitats, or may even make use of cutover secondary forest. However, there are many more species that are unable to adapt and that rely on pristine habitat for their continuing survival.

The world's most powerful bird of prey, the harpy eagle *Harpia harpyja,* is potentially threatened by the clearance and fragmentation of forest. This magnificent bird lives in the tropical forests of Central and South America, from Mexico to eastern Bolivia, southern Brazil, and extreme northern Argentina. It has been estimated that a pair of harpy eagles need a territory of undisturbed forest of 100 to 200 square kilometers (57 to 77 square miles) to satisfy their feeding requirements. Thus the minimum area of intact rainforest to ensure the harpy eagle's survival could be as large as 37,500 square kilometers (14,500 square miles) if it is assumed that a population of 250 pairs is necessary for perpetual viability. But as harpies tend not to overlap territory with their closest

Norman Tomalin/Bruce Coleman Ltd

competitor, the crowned eagle *Harpyhaliaetus coronatus* (already considered threatened), an even larger area may be necessary for the required 500 birds. At a very rough guess, 60,000 square kilometers (23,000 square miles) might be enough, an area equivalent to the tract of Amazonian forest destroyed in 1988. This species may only survive in the long term if the exponential rate of forest destruction is brought under control throughout its range, and a network of large undisturbed reserves is established.

Introduced species

Introduced species, especially predators such as cats, rats, and mongooses, have been the major cause of extinction of island birds. Many of these birds had evolved without any predator pressure and were unable to cope with introduced aliens which stole eggs and nestlings, or even hunted adults, especially of species that had lost the ability to fly. One such bird was the Stephen Island wren *Xenicus lyalli* (extinct 1894). Stephen Island is a tiny speck of land totalling 2.6 square kilometers (1 square mile) in the Cook Strait between the North and South Islands of New Zealand. The wren may have had the smallest natural range of any bird, as well as being the only passerine truly incapable of flight. But its main claim to fame was the remarkable manner of its extinction—by the lighthouse keeper's cat. This single feline discovered the species then destroyed the entire population in just a few months.

Introduced herbivores such as rabbits and goats can be as deadly as introduced predators because of the changes they wreak by browsing on vegetation. For example, the Laysan duck *Anas laysanensis* declined to near extinction on Laysan Island (a coral atoll belonging to the Hawaiian Islands) in the 1920s after the introduction of rabbits and the consequent denudation of the island's foliage. Two further species, the Laysan rail *Porzanula palmeri* (extinct on Laysan between 1912 and 1923, but persisting on Wake Island, where it was introduced, until 1944) and the Laysan millerbird *Acrocephalus familiaris* (extinct between 1912 and 1923) were unable to survive, whereas the Laysan finch *Telespyza cantans* was able to persist by changing its diet and feeding on seabird eggs.

Species introduced in the past still severely threaten a number of birds today. For example, the brown tree snake has been responsible for the population crash of all forest birds on the island of Guam in the Pacific, in particular the endemic Guam rail *Rallus owstoni* (extinct 1987, but surviving in captivity) and Guam flycatcher *Myiagra freycineti* (extinct 1985). Exotic fish in Madagascar have caused a decline in waterlilies and other aquatic plants, and so threaten the Madagascar little grebe *Tachybaptus pelzelni*. Nowadays efforts are made to keep islands free of introduced predators, and in some places

measures have been developed to eradicate the invaders. Cats had been present on Little Barrier Island, off Auckland, New Zealand, for over 100 years, but an eradication program between 1977 and 1980 resulted in their complete elimination and a marked improvement in a number of threatened bird species. For example, Cook's petrel *Pterodroma cooki* and the black petrel *Procellaria parkinsoni*, both heavily preyed on (100 percent of black petrel fledglings were killed from 1974 to 1975), have shown a significant recovery; the stitchbird *Notiomystis cincta*, once widespread in the forests of New Zealand but surviving only on Little Barrier Island, was declining but is now increasing; and the saddleback *Creadion carunculatus*, extinct on Little Barrier Island, has been successfully reintroduced.

Human predation

Hunting has been another major cause of extinction, and includes hunting for food (both birds and eggs), for feathers, and in some cases for

▲ *Introduced predators such as feral rats and cats have precipitated the extinction of many small birds. Once widespread in the original forests of New Zealand, the stitchbird shown here now survives only on a few tiny offshore islands, where constant vigilance is required to prevent the accidental introduction of predators.*

Brian Chudleigh

Hans Reinhard/Bruce Coleman Ltd

▲ *Red in the plumage is a quality that long eluded canary fanciers, until it was found that controlled cross-breeding with the red siskin of South America would introduce the necessary genetic material; as a consequence of this discovery, red siskins were trapped to the point that they are now an endangered species.*

1984–1985 the only record of the species was of one individual wintering at a Moroccan wetland. It seems likely that uncontrolled hunting over a long period of time has reduced the population year by year, and the breeding success (or lack of it) of the species has been unable to compensate for the hunting losses. If this species is to survive, a ban on all shooting of curlews and godwits (because of misidentification by hunters) in all these countries is urgently needed.

International trade

The wild-bird trade is a growing threat to birds. Some species are particularly sought after, and the rarer they become, the higher their market value. Thus it is very difficult to control the trade in prized birds because trappers and dealers are tempted by rich rewards. The red siskin *Carduelis cucullata* is one popular cage-bird. Trade in the species began in the nineteenth century, and demand escalated this century when it was discovered that it was possible to hybridize the red siskin with the domestic canary *Serinus canarius* (in the same subfamily of finches) and so introduce the genes for red plumage. The siskin once occupied a continuous range across northern Venezuela into northeastern Colombia, with isolated populations on the islands of Gasparee, Monos, and Trinidad. Thousands of birds have been trapped and numbers have declined drastically, so that now it is locally extinct throughout much of its former range, with total numbers in the high hundreds or low thousands.

Parrots are the most popular of all cage-birds, and the trade in some species has had devastating effects. Spix's macaw *Cyanopsitta spixii* has been illegally trapped down to the very last bird in the wild. Few ornithologists have ever seen the species and little is known about its way of life. It comes from a remote region of northeast Brazil, and in June and July 1990 a search for the bird was carried out over a large portion of its expected range. The survey team located only one bird. There were reports that Spix's macaw had persisted in another nearby location until the previous year, but residents thought that the last individuals had been taken by trappers. Further work indicated that the preferred habitat of the species was mature woodland along watercourses and that only three small patches currently remain in the state of Bahia. It is becoming clear that the range of this parrot is much smaller than previously thought, and the species is now virtually extinct because of trappers exploiting an already tiny population in a diminishing habitat. The only hope for the species lies with the 30 or so individuals held in captivity, and a carefully planned captive breeding program.

Pesticides and pollution

The use of pesticides and the pollution of the environment threaten many bird species

museum specimens. For example, the Hawaii mamo *Drepanis pacifica* (extinct 1899) was hunted for its brilliant yellow rump feathers, and 80,000 birds were sacrificed to make the famous royal cloak worn by Kamehameha I. Perhaps the most famous case of hunting to extinction is that of the passenger pigeon *Ectopistes migratorius* (extinct 1914). Once vast numbers darkened the North American skies, and hunting competitions were organized in which more than 30,000 dead birds were needed to claim a prize. The supply appeared inexhaustible, yet within a century the incredible multitudes were reduced to small dwindling bands, and in 1914 the sole survivor died in Cincinnati Zoo.

Many species are still hunted today, and while some are able to withstand the losses, others are seriously reduced. One species that has suffered particularly is the slender-billed curlew *Numenius tenuirostris,* one of the rarest and most poorly known species of the Western Palearctic region (Europe, Russia, Middle East, and North Africa). This species is believed to nest in Russia, and on migration it visits Ukraine, Turkey, Romania, Bulgaria, Hungary, the former Yugoslavia (Serbia), Greece, Italy, Tunisia, Morocco, and Iran. It appears to have been common in much of its range up to the end of the nineteenth century— observations from Algeria describe "incredible flocks ... as big as starling flocks"—yet in throughout the world. During the 1960s scientific proof emerged that more than 20 bird species across Europe and North America were suffering disastrous breeding failures because of malformed and broken eggs. The peregrine falcon *Falco peregrinus,* Eurasian sparrowhawk *Accipiter nisus,* golden eagle *Aquila chrysaetos,* osprey *Pandion haliaetus,* and brown pelican *Pelecanus occidentalis* were a few of the birds affected. These birds, at the top of their food chains, all showed high levels of DDT contamination, and this was linked beyond doubt to their unusually thin eggshells. Chlorinated hydrocarbons (of which DDT is one) persist in the environment for many years and, because they are soluble in fats and oils, accumulate in the fatty tissues of top predators and have a detrimental effect on their reproduction.

The use of chlorinated hydrocarbons is now restricted and the populations of affected species are recovering, but new classes of chemicals have taken their place and these too can have deadly side-effects. For instance, there is a general feeling that the use of pesticides in Africa may represent a serious threat to many European migrants (as well as to intra-African migrants and resident species). It is also claimed that locust and bird control campaigns—for example, of the quelea, which is a pest to agriculture—have caused casualties among an array of species of migratory birds including storks, herons, and birds of prey. Furthermore, there is circumstantial evidence that pesticides, especially rodenticides, are responsible for the decline of resident bird populations in Egypt. As well as pesticides and insecticides, there is a host of other synthetic chemicals which now pollute the atmosphere, the land, and the water, and which are likely to be harmful to birds (and to other wildlife and humans). High levels of polychlorinated biphenyls (PCBs) have been recorded in the carcasses of the white stork *Ciconia ciconia* in Israel and in addled eggs in Germany and Holland. These chemicals, produced by the combustion of plastics and other waste materials from the electronics industry, are known to severely reduce avian breeding success, and so could be a factor contributing to the decline of some species.

Climatic catastrophes

Changes in climate may also play a part in the decline of some bird species. For example, it has been suggested that the great auk *Alca impennis* of the North Atlantic (extinct 1844) was never very numerous and was restricted to a relatively narrow climatic zone. It seems likely that a period of severe weather coincided with increased human predation, and these combined factors eventually overwhelmed the species. But climatic change is not just a phenomenon of the past. We are currently experiencing a climatic change which may have far-reaching results—the so-called

greenhouse effect. If the use of fossil fuels such as coal, gas, and oil cannot be reduced, scientists speculate that the average global temperature could rise by 3°C (5°F) within the next 50 years. The world would be warmer than it has been for 10,000 years. Even this small change is likely to affect the ranges of many species, disrupt natural communities, and contribute to the extinction of some vulnerable species.

Some climatic changes can be sudden and dramatic. Hurricane Gilbert was the most powerful storm recorded in the Caribbean this century. When Gilbert hit Jamaica on September 12, 1988, gusts in excess of 220 kilometers per hour (137 miles per hour) were registered as winds tore across the island causing enormous destruction, especially to the montane and mid-level forests. It was feared that six endemic species largely confined to these habitats could have been badly affected. In fact the birds survived the hurricane surprisingly well. As hurricanes are not infrequent visitors to Jamaica, and the forests are probably always in some state of recovery, the avifauna is doubtless adapted to this dynamic system. But there is concern that continued forest clearance and charcoal extraction could reduce the forests to such tiny areas, and that hunting could reduce the bird populations to such low numbers, that next time they may not recover from the impact of a storm as fierce as Gilbert.

BIRDS WITH RESTRICTED RANGES

Some birds are considered threatened because they have tiny ranges—for example, birds confined to single small islands, like the Ascension frigatebird *Fregata aquila.* This seabird breeds only on Boatswainbird Islet (3 hectares, or 7 acres) just off the coast of the remote South Atlantic island of Ascension. It once bred on Ascension itself, but

▼ *Pollution has exacted its toll on wild bird populations. One of the effects of DDT on peregrine falcons is to interfere with the body chemistry needed by the female to properly form the shell around her egg. As a result, eggs are fragile and break easily in the normal stresses of incubation, killing the chick inside. This effect almost wiped out the peregrine, but with the banning of DDT, populations began a gradual recovery.*

R.T. Smith/Ardea London Ltd

41

► *An encouraging success story in recent bird conservation is the saving of the Lord Howe rail, confined to a tiny island in the Tasman Sea. Reduced to a critical 20 birds in the late 1970s, a captive breeding program and reintroduction program was instituted, and today the species numbers some 200 birds.*

Tom & Pam Gardner

▼ *Confined to tussock swamps in the most remote mountains of the South Island of New Zealand, the takahe was considered extinct until dramatically rediscovered in 1947. The species has been introduced to four island sanctuaries, but remains endangered.*

Brian Chudleigh

egg-collecting, disturbance, and the introduction of alien mammals, especially cats, have forced the frigatebird onto cat- and people-free Boatswainbird Islet. Despite the contraction of its range it still appears to be a fairly numerous species with a population of many thousands of birds. Nevertheless, the fact that the entire population is confined to such a minute area renders it a species that could be easily and quickly threatened should cats, for example, get a foothold on the island.

Many continental species also have restricted distributions, confined to single mountaintops or tiny patches of forest. Bannerman's turaco *Tauraco bannermani* and the banded wattle-eye *Platysteira laticincta* are two such species, found only in the shrinking forest of Mount Kilum (also known as Mount Oku) in the Bamenda Highlands of Cameroon. The black-breasted puffleg *Eriocnemis nigrivestis* is another species at risk because of its small range, being known only from two volcanoes in north-central Ecuador, and threatened by habitat destruction because of proximity to the capital city, Quito. It is vital that birds like these are given adequate protection before it is too late.

One restricted-range species that has been given the necessary protection is the noisy scrub-bird *Atrichornis clamosus,* a bird of dense scrub and forest in the extreme southwest of Western Australia. At one stage it was feared extinct, not being recorded since 1889, but in 1961 a small population of 40 to 45 singing males was rediscovered at Two Peoples Bay, east of Albany. The disappearance of the scrub-bird from most of its former range has been attributed to the intense fires started by European newcomers to encourage the growth of

grasses more suitable for cattle-grazing. The Two Peoples Bay population probably survived because the area was rugged and unsuitable for agriculture.

However, the rediscovery of the noisy scrub-bird presented a problem, for the area where the bird was living was precisely where the Western Australian government had proposed establishing a new town. To the government's credit the site was cancelled, and in 1966 Two Peoples Bay Nature Reserve was declared. In 1993 the population of singing males was estimated at 400 individuals spread over 30 kilometers (18 miles) of coastal and near-coastal land.

MYSTERY BIRDS

Some birds are included on lists of threatened species because they are virtual mysteries—they have been seen so few times that we assume they must be extremely rare and therefore threatened. The Fiji petrel *Pterodroma macgillivrayi* was formerly known from one specimen collected on Gau (an island in the Fiji group) in 1855. However, in 1984 an adult was captured on Gau and released, and in 1985 a fledgling was found there. So it is likely that the species breeds on the island, although there have been no further sightings. Predation by feral cats is a potential threat.

Mystery birds keep turning up. One recent discovery in 1988 was that of a new species of shrike from Somalia. The Bulo Burti bush-shrike *Laniarius liberatus* (described on the basis of blood and feather samples from a single individual) was found in dense *Acacia* bushland and had presumably escaped detection because of the impenetrable nature of its habitat. Another mystery bird is the Red Sea cliff swallow *Hirundo perdita,* known from one specimen found dead at a lighthouse off Port Sudan in 1984.

BACK FROM THE BRINK

Of the 1,000 or so species threatened with extinction there are some with so few individuals that they are on the very brink of extinction. And yet it may be possible to conserve even these species. The whooping crane *Grus americana* exemplifies such a species, being the center of a remarkable USA–Canadian conservation effort which has seen the total population rise from an all-time low of 15 wild birds in 1944 to over 340 birds in wild and captive flocks. The whooping crane has probably never been very numerous in historical times—possibly as few as 500 birds existed in the late eighteenth century. Species with naturally low numbers are especially vulnerable to change, but low numbers coupled with small (and shrinking) breeding and wintering grounds, and a hazardous (because of hunting and power lines) migration route over 3,000 kilometers (1,850 miles) linking the two, had stacked the odds against the chances of the whooping crane's survival. The story of the whooping crane recovery is a complicated one, involving much ferrying of eggs and birds from one location to another in a massive attempt to maximize the reproductive rate, but the enormous effort and expense has paid off.

The majority of threatened birds, however, live in developing countries where funding for conservation takes second place to the needs of people, and it is therefore unlikely that many species can be saved once they reach such a critical state. Instead it is vital that appropriate conservation action is taken at a much earlier stage.

Nothing more clearly illustrates the urgent need for swift action to save species on the brink of extinction than the story of Gurney's pitta *Pitta gurneyi,* a stunningly beautiful bird with brilliant blue crown and chrome-yellow flanks. The pitta is a resident of southern Myanmar (Burma) and peninsular Thailand, and though common at the start of this century it had not been encountered in the wild since 1952. Much of its original lowland forest habitat had been cleared and settled by a burgeoning human population. A search in 1984 and 1985 revealed no birds, but in 1986, following a lead from a bird dealer, the quest for the species was rewarded with the discovery of a nesting pair in a tiny patch of forest at Khao Noi Chuchi in the far south of Thailand. Subsequent intensive survey work has shown that virtually the entire population of some 30 pairs is confined to this one forest. Khao Noi Chuchi was destined to be logged and encroachment by local people had already started, but a petition to the Minister of Agriculture drawing attention to this priceless forest fragment (apart from the pitta, it harbors a spectacularly diverse range of other animals and plants) brought some protection. Now a major conservation project is helping to protect the forest and at the same time to aid the nearby farmers so they have no need to destroy the forest. Without this action Gurney's pitta would almost certainly be extinct today.

CONSERVATION IN THE FUTURE

The species of birds at risk of extinction, and those that are potentially threatened or declining, must be saved. Different species have different conservation needs, but some combination of preserving habitat, eliminating introduced species, banning hunting, controlling trade, and preventing pollution will help to ensure their survival. But assessing the required action and executing it takes time and money, and both are in short supply. While conservation action to save flagship species like Gurney's pitta can be successful—not just for the species itself but also for associated wildlife—conservation in the future is best focused on key sites where several threatened species occur together. For it is in these "hotspots" that there will be the greatest return for the conservation effort invested. If we preserve such sites we will make a major contribution to maintaining the earth's biological diversity.

ALISON STATTERSFIELD

Brian Chudleigh

▲ *The black robin was snatched from the brink of extinction by determined last-minute measures. Confined to Little Mangere Island in the Chatham Islands in the far southwestern Pacific, it was reduced to just one viable pair before a complex cross-fostering program began to offer some hope that its numbers might be rebuilt. Today it numbers more than 150 birds.*

▼ *Generally, birds of lowland tropical rainforest seem most at risk: this is the habitat of the critically endangered and beautiful Gurney's pitta.*

M.D. England/Ardea Photographics

PART TWO

KINDS OF BIRDS

RATITES & TINAMOUS

Ratites, the giants among birds, and their small relatives, the tinamous, are the living representatives of the order Struthioniformes, though the tinamous are sometimes placed in a separate order. All species are restricted to southern continents, which has encouraged the view that they represent a primitive group that did not penetrate the Northern Hemisphere. Recent discoveries of ratite fossils in Europe, however, indicate that they may have once been wide ranging. Members of this order—the ostrich, rheas, cassowaries, emu, and kiwis—show that birds can evolve into large flightless vertebrates comparable with the large herbivorous mammals.

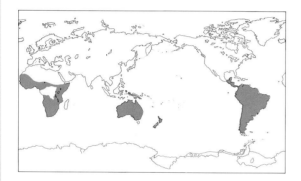

AN UNKEELED BREASTBONE

The order contains six living families (or five if the emu is placed with the cassowaries): Struthionidae, represented by the ostrich of Africa, the world's largest bird; Tinamidae, about 45 species of tinamous in South and Central America; Rheidae, two species of rhea in South America; Casuariidae, the cassowaries, three species in New Guinea and Australia; Dromaiidae, the emu of Australia; and Apterygidae, three species of kiwi in New Zealand. In the historical past the Diornithidae, with at least 12 species of moas, inhabited New Zealand, and individuals may have survived into the nineteenth century. The final date of extinction of another family, the Aepyornithidae (elephantbirds), confined to Madagascar, was about 1650.

All "true" ratites are flightless, and they have a flat sternum (breastbone) without the keellike prominence of most flying birds. (The Latin word ratis means "raft"). The tinamous have a keeled sternum, and also have the ability to fly, but they resemble the other Struthioniformes in many unusual anatomical characters such as the structure of the palate. Their plumage is loose compared with the feathers of most other birds. Cassowaries and the emu have plumage that hangs like hair from their bodies; each feather has two shafts of equal length. The ostrich, rheas, tinamous, and kiwis have feathers with one main shaft, but the barbules, if present, do not interlock closely so the birds appear shaggy.

THE OSTRICH: A KING WITH HIS HAREM

Ratites are running birds. They gain a mechanical advantage in having long, thin legs to support the body's weight well above the ground, in a similar way to the ungulates (horses, cows, and their relatives). The number of toes has been reduced in the course of ratite evolution. Most birds have four toes—in ratites, the kiwis, moas, and some tinamous have four toes; the emu, rheas, and many tinamous have three toes; and the ostrich has only two toes. In the ostrich, unlike in most ratites, males are larger than females; males grow to 2.75 meters (9 feet) tall, females to 1.9 meters (6¼ feet) tall. The ostrich's wide range once included the Middle East, North Africa, and Africa south of the tropical rainforests, but it is now extinct in the Middle East and most of North Africa. Many southern African populations are confined to national parks.

Ostriches use their large wings in courtship, and each cock builds a nest to which he attracts a hen. She lays her eggs there and becomes the major hen. Other hens (minor hens) also lay eggs there, but only the major hen and the cock incubate. She selectively keeps her own eggs in the nest, discarding some of those laid by other hens. Eventually 60 or more eggs may be laid in or around the nest but only about 20 are incubated. The major hen incubates by day and the cock takes over at night during the incubation period of 39 to 42 days. Both sexes guard the chicks, which may remain as a family for 12 months. Nests are usually spaced about 2 kilometers (1¼ miles) apart, but in dense bush they may be closer than that, and on the plains, they may be more widely spaced.

Large concentrations of ostriches occur daily around water or where food is abundant, and immatures are found in flocks of up to 100 birds. The ostrich's diet is a selection of fruits, seeds, succulent leaves, and the growing parts of shrubs, herbs, and grasses. They also take small vertebrates.

◄ *The ostrich was once widespread across Africa and the Middle East, but it is now essentially confined to various national parks in Africa. Until about the turn of the century the ostrich was widely farmed for its feathers, which were used in the fashion industry; the industry is not entirely defunct, and captive flocks in South Africa still total about 60,000 birds. In the Arab world the shells of ostrich eggs are credited with magical powers, and are often used on the roofs of Muslim homes or Egyptian Orthodox churches with the intention of deflecting lightning.*

THE SOUTH AMERICAN FAMILIES

The two representative families of ratites in South America, the rheas and the tinamous, total 47 species and inhabit a variety of habitats from forest, to the high antiplano (puna) of the Andes.

Rheas

Rheas are sometimes called South American ostriches. The greater rhea *Rhea americana* stands about 1.5 meters (5 feet) tall and weighs 20 to 25 kilograms (44 to 55 pounds). The lesser rhea *Pterocnemia pennata* is smaller. Both species have gray or gray-brown plumage, with large wings that cover the body like a cloak. When they run rheas sometimes spread their wings, which then act as sails, but the birds are unable to fly.

The original distribution of the two species was unusual. The greater rhea lived on the plains from northeastern Brazil to central Argentina, but its range has been dissected by agricultural

Michael Yamashita

▲ A one-wattled cassowary. Protein being scarce for many native peoples in the rainforests of New Guinea, a common practice is the capture and raising of young cassowaries, which are slaughtered for food when they reach maturity.

▼ The tinamous of South America are in many respects the ecological equivalents of quail and partridges elsewhere in the world. All are terrestrial, though many roost in trees. Some species, like this highland tinamou shown at its nest, inhabit rainforest.

Michael Fogden/Bruce Coleman Ltd

development. The lesser rhea has two separate populations: one on the pampas (grasslands) of Patagonia, known as Darwin's rhea; and the other in the high Andes of southern Peru and northern Chile, the puna rhea.

Rhea males fight for territories, and once a male has established his domain he builds a nest, a scrape on the ground lined with leaves and grass. To this he attracts females, often as a small flock. Each female lays an egg in the nest, returning to do so every two or three days until the male, responding to the size of the clutch, drives them away and begins to incubate. Before and after the females lay in the nest, however, they lay eggs on the ground in the vicinity, some of which the male rolls beneath him; the others rot. Once he sits, the female flock goes off to attend another male and may serve half a dozen nests in a season. The cock leads the chicks, which grow quickly and are of adult size in about six months but do not breed until they are two years old.

Tinamous

The 45 species of tinamou vary from the size of a quail to that of a large domestic fowl. While they show close relationships to other Struthioniformes in their anatomy, their eggwhite proteins, and the structure of their genes, they differ in some conspicuous ways. Many, perhaps all, species can fly, although they seldom do so; they usually escape predators by stealing away through cover or freezing. Most species have three toes, a common ratite number, but some have four. Members of the genus *Tinamus* roost in trees; all other tinamous roost on the ground. In *Tinamus* species the back of the tarsus is roughened to give the birds a good grip on the branch when at rest.

Many species feed on vegetable matter, but some (for example, *Nothoprocta* species) take much

animal food, and the red-winged tinamou *Rhynchotus rufescens* digs for roots and termites. In some species (such as the ornate tinamou *Nothoprocta ornata*) a single male and female establish the nest, but in most a male associates with several females at nesting. Tinamous nest on the ground, lining a depression with grass and leaves, and like most ratites the male undertakes the incubation (19 days for *Eudromia*) and looks after the family.

Tinamous are found in many habitats, including rainforest and the high and barren Andes. On the open tablelands the martineta tinamou *Eudromia elegans* lives in flocks of up to a hundred. Tinamous are diverse and abundant, an impressive achievement considering that they are now thought likely to be close to the ancestral stock of all ratites.

EMU AND CASSOWARIES

The Australian emu *Dromaius novaehollandiae* lives a nomadic existence, continually moving to keep in touch with its food—not that the food moves, but rather that abundances of flowers, fruits, seeds, insects, and the young shoots on which it feeds appear in random sequence in the Australian deserts. Emus, standing 2 meters (6½ feet) tall and weighing up to 45 kilograms (100 pounds), move over vast distances, usually as monogamous pairs, stopping when they find abundant food and moving again when it is exhausted. Only when the male undertakes the eight weeks of incubation is it impossible for him to move to find food. During incubation he does not eat, drink, or defecate, living instead on the fat reserves he has built up in the previous six months. If conditions have not allowed the pair to store fat before the winter breeding season, they do not breed, or if eggs are laid the male may desert them before they hatch. The male guards the chicks and leads them for their first seven months and sometimes longer. The female may remain nearby, or move far away in search of food, or mate with another male. Seldom do the pair re-form for a second season.

Emus live throughout southern Australia, not just in deserts. They become less common in the north, although a few birds can be found as far north as Darwin and Cape York. Their numbers, currently estimated at about 500,000, can rise or fall rapidly in correlation with wet and dry seasons. They are common in coastal scrub, in eucalypt woodland, and on saltbush plains. A few venture into alpine heath, and many are still present in farming areas provided some bushland remains.

Cassowaries favor jungle. Three species live in New Guinea, and one of these, the double-wattled cassowary *Casuarius casuarius,* also lives in the tropical rainforests of far northeastern Queensland, Australia. In New Guinea the original distribution of the three species is uncertain because humans have transported them and released them beyond their natural range. It is likely that the double-

C.A. Henley

wattled cassowary favored mid-level rainforest, the one-wattled cassowary *C. unappendiculatus* low-level rainforest, and the dwarf cassowary (or moruk) *C. bennetti* the highlands, perhaps even the montane grasslands. Several islands around New Guinea have cassowary populations, usually of one species—for example, the double-wattled cassowary on Ceram and the moruk on New Britain. The double-wattled cassowary stands 1.5 meters (5 feet) tall and may weigh more than 55 kilograms (120 pounds). Its glossy black plumage grows after the first year; before that the young birds are clad in a sober gray. All species have throat wattles, brilliant red and blue in adults but less colorful in immatures. The moruk does not have a distinct casque, but a casque adorns the heads of the other two species. Cassowaries depend on forest fruits for food—fruits from more than 75 species of tree in northern Queensland. Individuals seem to maintain a territory of 1 to 5 square kilometers (⅓ to 2 square miles), moving around it to gather fruit as different trees ripen. The territory is occupied by pairs during the winter breeding season, and the clutch of six to eight eggs is incubated for about two months by the male, who also looks after the young chicks. Cassowaries are not abundant anywhere, and their survival will be imperiled if the diversity of the forests in which they live is reduced by logging.

KIWIS, THE BURROWING RATITES
Three species of kiwi remain in New Zealand: the brown kiwi *Apteryx australis* is the largest, 55 centimeters (21 inches) long, with females

weighing 3.5 kilograms (7¾ pounds); the great spotted kiwi *A. haastii*, intermediate in size; and the little spotted kiwi *A. owenii*, 35 centimeters (14 inches) long, weighing 1.2 kilograms (2½ pounds). The brown kiwi is still found on North, South and Stewart Islands, the great spotted only on South Island, and the little spotted on Kapiti Island, where it was introduced in 1913.

Kiwis are nocturnal, feeding on invertebrates which they find mainly by scent, probing with their long and sensitive bills. Pairs form during the late winter/spring breeding season, and the male excavates a burrow in which the female deposits one to three white eggs. Each egg is equivalent to 25 percent of the female's body weight, proportionately the largest egg laid by any bird. The eggs are incubated by the male for 78 to 82 days, and the chicks appear to be independent almost from the time they emerge from the burrow.

The call of the male brown kiwi is a shrill whistle with a long ascending phrase and a short descending one at the end giving rise to the name "kiwi". Females have a hoarse, low cry. Two anatomical features set kiwis apart from other ratites. Firstly, the wings—small in the emu and cassowaries—are vestigial in kiwis. Secondly, female kiwis have paired, functional ovaries—in most birds usually only the left is functional, and the right is absent altogether. All species live in native podocarp (southern conifer) forest, but the brown kiwi has survived in farmland and in pine forest, although it is still unclear if such populations are self-sustaining outside these natural forests.

S.J.J.F. DAVIES

▲ *Emus sometimes form wandering parties when not breeding, but after mating and the eggs are laid, the male incubates and raises his brood of young alone. The chicks are covered with down when hatched, and can follow their father within a few hours. He indicates food to his chicks, but they normally feed themselves.*

▼ *The kiwi is New Zealand's national symbol, though in fact there are three very similar species. Restricted to New Zealand, the group has no close relatives anywhere in the world. All three kiwis are flightless, nocturnal, and live in burrows. The great spotted kiwi, shown here, occurs only on the South Island.*

Brian Chudleigh/National Wildlife Centre

ORDER PROCELLARIIFORMES
- 4 families • 23 genera
- 92 species

SMALLEST & LARGEST

Least storm petrel *Halocyptena microsoma*
Body length: 12½–15 cm (5–6 in)
Wingspan: 32 cm (12½ in)
Weight: 28–34 g (1–1⅕ oz)

Wandering albatross *Diomedea exulans*
Body length: 1.1–1.4 m (3⅗–4⅗ ft)
Wingspan: 3.4 m (11 ft)
Weight: 6–11 kg (13–24¼ lb)

CONSERVATION WATCH
!!! There are 11 species listed as critically endangered: Amsterdam albatross *Diomedea amsterdamensis*; Mascarene black petrel *Pterodroma aterrima*; Chatham Islands petrel *Pterodroma axillaris*; Barau's petrel *Pterodroma baraui*; Beck's petrel *Pterodroma becki*; Jamaica petrel *Pterodroma caribbaea*; Fiji petrel *Pterodroma macgillivrayi*; Zino's petrel *Pterodroma madeira*; magenta petrel *Pterodroma magentae*; Galapagos petrel *Pterodroma phaeopygia*; Guadalupe storm petrel *Oceanodroma macrodactyla*.
!! The 6 species listed as endangered are: short-tailed albatross *Diomedea albatrus*; Bermuda petrel *Pterodroma cahow*; black-capped petrel *Pterodroma hasitata*; Heinroth's shearwater *Puffinus heinrothi*; Hutton's shearwater *Puffinus huttoni*; Peruvian diving petrel *Pelecanoides garnotti*.
! 15 species are listed as vulnerable.

ALBATROSSES & PETRELS

Birds of the order Procellariiformes, known collectively as petrels or tubenoses, are highly adapted to a marine way of life, spending much of their time at sea and feeding on the larger zooplankton, cephalopods (squid), and fish. They can be found throughout the world's oceans, but it is thought that their ancestors evolved in the southern seas. Their fossil record goes back some 60 million years.

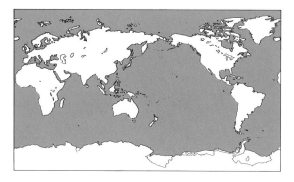

BIRDS OF THE OPEN OCEAN
A number of characteristics are shared by the Procellariiformes, notably their long, external, tubular nostrils — hence the vernacular name, tubenoses. This special feature is associated with a comparatively (for birds) well-developed olfactory part of the brain, and it is likely that the Procellariiformes locate their food, breeding sites, and each other by their sense of smell. All species have a distinctive musty body odor, which persists for decades in museum specimens.

Other characteristics include a deeply-grooved, markedly-hooked bill, and the ability of most species (not the Pelecanoididae, diving petrels) to store large quantities of oil in their stomachs. This oil, derived from the energy-rich food stores laid down by the marine organisms on which they feed, is regurgitated to feed the young or ejected to deter would-be predators. It is one of a number of adaptations for the highly pelagic life, with vast oceanic ranges, typical of the group. Recent satellite tracking studies have shown that the wandering albatross *Diomedea exulans* covers between 3,600 and 15,000 kilometers

R. Drummond

▶ *The wedge-tailed shearwater inhabits tropical and subtropical waters of the southern Pacific and Indian oceans.*

Graham Robertson/Auscape International

(2,200 to 9,300 miles), flying at speeds of up to 80 kilometers (50 miles) per hour, in a single trip while its partner takes a 10-day incubation shift.

The Procellariiformes can be divided into four families: albatrosses (family Diomedeidae), shearwaters (family Procellariidae), storm petrels (family Hydrobatiidae), and diving petrels (family Pelecanoididae).

ALBATROSSES

When fifteenth-century Portuguese navigators first ventured down the coast of Africa into the windy south Atlantic they encountered large black-and-white seabirds with stout bodies and long pointed wings. They called these strange birds *alcatraz,* the Portuguese word for large seabirds. English sailors later corrupted alcatraz to albatross.

Albatrosses are the supreme exponents of gliding flight, sometimes gliding for hours without a single wing-beat. They are typically associated with the belt of windswept ocean lying between the Antarctic and the southern extremities of America, Africa, and Australia, but they also breed in the more temperate waters of the Southern Hemisphere. Three species breed in the north Pacific, and one species breeds at the Galapagos Islands on the Equator.

Albatrosses are distinguished from the other Procellariiformes by the position of their tubular nostrils, which lie at either side of the base of the bill, rather than being fused on the top of the bill. They eat a variety of food, often scavenging behind ships, but fish, squid, and crustaceans are favorite items and are frequently caught at night. They catch prey mainly from the surface of the sea, but occasionally catch it from just beneath the water by plunge-diving

into the waves with bent wings.

Albatrosses are long-lived birds with an average life span of 30 years. But what they gain in longevity they lose in productivity, with most species not breeding until they are 10 to 15 years old, and several species breeding only biennially. Most albatrosses breed in closely-packed colonies, sometimes numbering thousands of pairs, but sooty albatrosses of the genus *Phoebetria* are solitary or nest in small groups on cliff edges. For several species the nest is a heap of soil and vegetation, although the tropical species make do with a scanty nest and the waved albatross *Diomedea irrorata* doesn't prepare one at all.

Albatrosses usually pair for life, and "divorce" occurs only after several breeding failures. They are famous for their spectacular courtship displays, standing with wings open and tails fanned, while stretching out their necks and throwing back their heads until the tip of the bill is buried in the plumage of the back. All this is accompanied by gurgling and braying sounds, and in the case of the great albatrosses, which have the most complex rituals, by the rattling of their bills.

In common with all Procellariiformes, albatrosses lay a single white egg which is incubated in alternate shifts of several days by both parents, from about 65 days in the smaller species to 79 days in the royal albatross *Diomedea epomophora.* The chick is initially brooded for three to four weeks, but thereafter the parents remain ashore only long enough to transfer a meal. Fledging takes from 120 days for the black-browed albatross *D. melanophrys* and yellow-nosed albatross *D. chlororhynchos,* to 278 days for the wandering albatross. The extremely long nesting period of the latter species, and of five others (royal

▲ The wandering albatross differs from most other tubenoses in the bewildering series of different immature plumages worn by each individual through its lengthy adolescence. Old males have largely white upper wings, unlike these rather young birds at their breeding ground on Macquarie Island. Albatrosses mate for life, and greet each other on their return at nesting time with spectacular displays involving outstretched wings and trumpeting calls.

Graham Robertson

▲ *The yellow-nosed albatross is typical of the mollymawks, a group of rather small albatrosses with plain dark upperwings and colorful bills. Like all albatrosses it is a bird of the open ocean, seldom approaching land except to nest. This species breeds annually (not in alternate years like the great albatrosses) on several islands in the Indian and southern Atlantic oceans.*

Mollymawks

This group includes nine smaller species of albatross with an average wingspan of 2.2 meters (7 feet). The term mollymawk comes from the Dutch word *mollemok* meaning "foolish gull". Three species breed in the north Pacific: the short-tailed or Steller's albatross *Diomedea albatrus,* from the tiny volcanic island of Torishima off Japan; the black-footed albatross *D. nigripes,* from the northwest Pacific; and the Laysan albatross *D. immutabilis,* from the Hawaiian archipelago. The waved albatross is the only equatorial species of albatross, breeding on the Galapagos and Isla de la Plata off Ecuador, and wintering in the rich waters of the cool Humboldt Current off Ecuador and Peru.

Sooty albatrosses

These are similar in size to the mollymawks and include two species only: the light-mantled sooty albatross and the sooty albatross. Both species breed on temperate and subantarctic islands in the Atlantic and Indian oceans, south and north respectively of the Antarctic Polar Front, and co-occur at a few sites. They are dark-plumaged birds with more slender wings and longer pointed tails than other albatrosses.

SHEARWATERS

The shearwaters are the most diverse family of the Procellariiformes, ranging in size from the giant petrels (the size of albatrosses) to diminutive prions, but typical members are about the size of small gulls. They get their name from some species' habit of skimming just over the surface of the sea. They have the widest distribution of any bird family, with species nesting 250 kilometers (155 miles) inland in Antarctica (the snow petrel *Pagodroma nivea*) and as far north as there is land in the Arctic (the northern fulmar *Fulmarus glacialis*). Despite this great variation, the family can be divided into four groups which have some common attributes.

Fulmars

The fulmars are a cold-water group, only venturing into the subtropics along cold-water currents. All of them have large, stout bills and feed on larger zooplankton and fish; most are proficient scavengers and characteristically associate with fishing vessels. Of the seven species, six are confined to the Southern Hemisphere. The single northern representative is the northern fulmar, which has shown a remarkable recent increase in range and numbers. Its southern counterpart is the Antarctic fulmar *Fulmarus glacialoides.* Three species of fulmar belong to genera with only one species in each genus: the Antarctic petrel *Thalassoica antarctica,* the Cape petrel *Daption capense,* and the only small all-white petrel, the snow petrel.

In contrast to these medium-sized species, which have an average wingspan of 1 meter, or 3¼ feet, the northern giant petrel *Macronectes halli*

albatross, Amsterdam albatross *D. amsterdamensis,* gray-headed albatross *D. chrysostoma,* light-mantled sooty albatross *Phoebetria palpebrata,* and sooty albatross *P. fusca*) means that these species, if they are successful in rearing a chick, breed only in alternate years.

The 14 albatross species can be conveniently divided into three groups.

Great albatrosses

The great albatrosses are at once recognizable by their huge size, the average wingspan being about 3 meters (10 feet). There are three species: the wandering albatross (breeding at most subantarctic islands) and the royal albatross, which are both circumpolar in the southern oceans, although the royal albatross is far less common away from its breeding islands around New Zealand; and the Amsterdam albatross, from the French subantarctic island of Amsterdam in the Indian Ocean, a recently described and nearly extinct species, with about five pairs breeding each year.

and southern giant petrel *M. giganteus* (both wide-ranging in southern oceans) are heavier than some albatrosses though have a wingspan of only 2 meters (6½ feet). Giant petrels take a wide variety of prey, but are notable as specialist scavengers at sea and ashore, where they exhibit vulture-like behavior at seal carcasses.

Prions

Prions are another southern group of petrels breeding chiefly on subantarctic islands. They are small (wingspan about 60 centimeters, or 2 feet) and are all very similar in size and appearance, blue-gray above and white below with distinctive W markings across their wings. They were once known as whalebirds because they were frequently seen in the presence of whales. There are about six species of prions (the status of some subspecies is still uncertain), which form two groups. Four species excavate underground burrows and have comb-like outgrowths from the palate to filter surface zooplankton from seawater. The other two species, the fulmar *Pachyptila crassirostris* and the fairy prion *P. turtur,* often nest among boulders, are less widespread (breeding mainly in Australian waters), and have less specialized bills.

Gadfly petrels

The gadfly petrels are a difficult group to classify and identify. Many species are very similar in coloration and markings, and there is controversy as to the number of distinct species (about 30). There is also debate as to whether the blue petrel *Halobaena caerulea* belongs to this group or to the prions. Gadfly petrels mostly nest underground in large colonies, but many species breed on remote islands where predators are absent.

They are medium-sized birds (wingspan about 80 centimeters, or 2½ feet), whose bills are short and stout with a sharp cutting edge and strong hook for gripping and cutting up small squid and fish. They are widely distributed in tropical and subtropical seas, particularly in the Pacific. Some have a very restricted distribution and breed on single islands, such as the cahow *Pterodroma cahow* from Bermuda, and the freira *P. madeira* from Madeira. Many species are very poorly known: the Mascarene petrel *P. aterrima,* for example, known only from Réunion by four specimens collected in the nineteenth century and three birds found dead in the 1970s; the magenta petrel *P. magentae,* recently rediscovered on a small stack off the coast of Chatham Island, New Zealand, 111 years after the unique type specimen was collected in the south Pacific Ocean; and Jouanin's petrel *Bulweria fallax,* described only in 1955, its breeding grounds still remaining unknown.

True shearwaters

The fourth group of shearwaters are the true shearwaters of the genera *Procellaria, Calonectris,* and *Puffinus,* about 23 species in total. They vary in size, with a wingspan from 60 centimeters (2 feet) to 1.5 meters (5 feet). They are widespread and very mobile. Shearwater bills (except *Procellaria*) are proportionately longer and thinner than those of fulmars, prions, or gadfly petrels, and are mostly used for seizing fish or large zooplankton underwater. The four *Procellaria* species are the largest members of the group and are restricted to the southern oceans, although the black petrel *P. parkinsoni* is known to winter off western Mexico. The two *Calonectris* species are geographically isolated, Cory's shearwater *C. diomedea* breeding in the north Atlantic and the streaked shearwater *C. leucomelas* breeding in the north Pacific; both migrate south during northern winters. The genus *Puffinus* comprises about 17 species of small to medium-sized shearwaters; the short-tailed shearwater *P. tenuirostris,* which breeds off southern Australia, is sometimes referred to as the muttonbird because chicks are harvested commercially for meat and oil. Most shearwaters nest in vast colonies of burrows which they visit at night; they are highly social, often forming dense feeding "rafts" at sea. Several of the southern species are transequatorial migrants, wintering in the north Pacific.

STORM PETRELS

Storm petrels are small birds (average wingspan 45 centimeters, or 1½ feet) and are found in all oceans. Many species have striking black plumage

Graham Robertson

▲ *A sooty albatross at its nest on Gough Island in the southern Atlantic Ocean.*

▼ *The tubenoses are most strongly represented in the southern hemisphere, but the northern fulmar is common in colder parts of the northern Atlantic and Pacific oceans, where it has increased dramatically in range and number this century. Fulmars are scavengers, and often congregate around fishing boats for offal.*

Jean-Paul Ferrero/Auscape International

▶ *Diving petrels use their wings more for propulsion underwater than for flight in air; their bodies are somewhat tubby and their wings small and rigid. In their general appearance and behavior they show some remarkable parallels with the auks , which do not occur in the Southern Hemisphere. This is the common diving petrel, the most widespread of the four species.*

Graham Robertson

with a white rump, and all are colonial and nest underground, mostly on isolated islands. Long migrations are often undertaken. For example, Wilson's storm petrel *Oceanites oceanicus* migrates from its Antarctic breeding grounds to the subarctic oceans of the Northern Hemisphere, and Matsudaira's storm petrel *Oceanodroma matsudairae* appears to migrate along a west–east axis from the Pacific to the Indian Ocean.

The northern group of storm petrels, about 13 species in total, have pointed wings and short legs. They swoop down and pluck food from the surface of the sea. The 11 species of the genus *Oceanodroma* have moderately forked tails.

The seven species of southern storm petrels are characterized by short, rounded wings and long legs, which are often held down as the birds bounce or "walk" on the sea surface while feeding.

DIVING PETRELS

As their name implies, diving petrels dive for their food, simply flying into the water and disappearing, and then suddenly reappearing from the depths, traveling through air and water with equal ease. Their wings are small and broad, an adaptation for swimming underwater, and their flight is whirring rather than gliding.

Their tubular nostrils open upwards rather than forwards, which is presumably an adaptation to their diving habit. All four species, which belong to one genus, *Pelecanoides,* are found in the southern oceans. They are small, with an average wingspan of 30 centimeters (1 foot). Unlike many other petrels, diving petrels do not appear to make extensive movements even outside the breeding season and are usually seen in the waters near their breeding areas.

THREATS TO SURVIVAL

Traditionally the albatross was thought to be a bird of ill omen, embodying the souls of drowned sailors. To kill one was bad luck. Despite this superstition, made famous by Coleridge's poem *Rime of the Ancient Mariner,* the albatross hasn't escaped persecution, being caught by many seafarers to relieve the monotony of their diets on long voyages.

Discovery of the remote island breeding sites favored by albatrosses and other Procellariiformes has led to losses through egg-collecting, and almost wholesale slaughter at some colonies for their luxuriant feathers, much in demand as soft bedding and popular with ladies' fashions at the turn of the century. The short-tailed albatross suffered particularly at the hands of the plume-hunters. Once abundant in the north Pacific, by 1953 it was reduced to about 10 pairs on the volcanic island of Torishima (eruptions in 1902 and 1939 also contributed to its near-extinction). The population is slowly recovering, with 146 adults and 77 fledglings recorded in 1986, and breeding success has improved since grass transplantation to stabilize nesting areas on the precarious volcanic slopes.

The biggest threats today are from predators such as cats and dogs, introduced by humans to the remote islands where petrels breed, and from commercial fisheries. In addition to competition for food, there is increasing evidence of substantial mortality from gill and drift-net fisheries, which mainly affect shearwaters, and from long-line tuna fisheries, which mainly involve albatrosses. This is particularly serious for birds such as Procellariiformes which have low reproductive rates.

JOHN P. CROXALL

PENGUINS

Penguins form a distinct group of highly specialized, social, flightless pelagic seabirds, widely distributed throughout the cooler waters of the southern oceans. The greatest concentrations and largest number of species occur in the sub-antarctic between latitudes 45° and 60°S, with the greatest diversity of species in the New Zealand area and around the Falkland Islands. Only two species are restricted to south of latitude 60°S in the Antarctic. Penguins are absent from the Northern Hemisphere, although the Galapagos penguin *Spheniscus mendiculus* sometimes ranges slightly north of the Equator. Other species inhabit the mainland coasts and offshore islands of Australia, New Zealand, Patagonia, Tierra del Fuego extending northward to Peru, and offshore islands of southwestern Africa. Although most penguins inhabit regions free of terrestrial predators, at sea they must contend with such efficient aquatic predators as carnivorous leopard seals and killer whales. In some areas, skilled aerial predators such as skuas take substantial numbers of chicks and eggs.

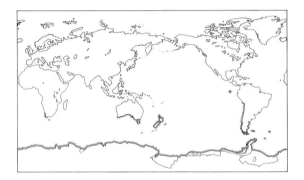

UNDERWATER SWIMMERS

Penguins have remained essentially unchanged for at least 45 million years. Although flightless, they evolved from flying birds. The most aquatic of all birds, some species may spend up to three-quarters of their life in the sea, coming ashore only to breed and molt. Many are migratory. With wings that have evolved into stiffened, flattened, paddle-like flippers, penguins are supreme swimmers. The only other birds that swim underwater using their wings rather than feet are auks and their allies (the Northern Hemisphere counterparts of penguins) and diving petrels. While penguins can attain speeds of up to 24 kilometers per hour (15 miles per hour) during brief stints, they generally swim at 5 to 10 kilometers per hour (3 to 6 miles per hour). They are by far the most accomplished of avian divers. The emperor penguin *Aptenodytes forsteri* can dive to 540 meters (1,772 feet) and remain submerged

KEY FACTS

ORDER SPHENISCIFORMES
• 1 family • 6 genera • 17 species

SMALLEST & LARGEST

Little blue or fairy penguin
Eudyptula minor
Total length: 39–41 cm (15–16 in)
Weight: 0.7–2.1 kg (1½–4½ lb)

Emperor penguin *Aptenodytes forsteri*
Total length: 115 cm (45 in)
Weight: 19–46 kg (42–100 lb)

CONSERVATION WATCH
! There are 5 vulnerable species: Fiordland penguin *Eudyptes pachyrhynchus*; Snares Islands penguin *Eudyptes robustus*; erect-crested penguin *Eudyptes sclateri*; yellow-eyed penguin *Megadyptes antipodes*; Galapagos penguin *Spheniscus mendiculus*.

▼ *Penguins are often popularly associated with the icy wastes of Antarctica, but the Galapagos penguin is restricted to an archipelago lying almost exactly on the Equator.*

for more than 20 minutes. Even much smaller species such as the gentoo penguin *Pygoscelis papua* dive to depths exceeding 150 meters (500 feet). Penguins feed on fish, krill, and other small invertebrates, and cephalods (squid), which are captured and consumed underwater.

All penguins are faced with thermoregulatory challenges: the polar penguins must conserve heat, whereas the temperate and tropical species have to shed excess heat. Thus the well-insulated south polar penguins generally have relatively smaller appendages, and feathering may extend well down on the bill. Conversely the tropical penguins have larger appendages and bare skin about the face which, when flushed, provides a mechanism for dissipating excess heat. No other group of birds is forced to endure air temperatures ranging from -60°C (-75°F) during the dark Antarctic winter to more than 40°C (105°F) at the Equator. Penguins depend on the insulative quality of their overlapping feathers to maintain their body temperature, the dense waterproof layer effectively trapping warm air. During the annual molt, when all feathers are lost simultaneously, a penguin is not waterproof and must come ashore or onto the ice. Molting birds cannot enter the sea to feed and therefore during the molt period of three to six weeks the fasting birds may lose a third or more of their body weight.

Like most seabirds, penguins tend to be rather long-lived, although juvenile mortality may be high. In the breeding season most species are highly territorial, but the emperor penguin forms large "huddles" during the winter. Some species do not become accomplished breeders until their tenth year. Upon hatching, the chicks are down-covered but are dependent on the adults for warmth and protection. Chicks are fed via regurgitation, and in surface colonies the parents recognize their young by voice.

THE SIX GENERA

The two largest and most colorful species, the king and emperor penguins, are both included in the genus *Aptenodytes*. Unlike other penguins which typically produce two-egg clutches, both lay a single egg which is incubated on top of the feet and covered by a muscular fold of abdominal skin. The emperor penguin is unique in that it breeds during the height of the dark Antarctic winter; only the males incubate for the entire incubation period of 62 to 67 days; and colonies are typically located on the annual fast ice, thus the emperor penguin is the only bird (under normal conditions) never to set foot on solid ground. The fasting period of 110 to 115 days endured by incubating males is the longest for any bird.

The smaller but more colorful king penguin *A. patagonicus,* of subantarctic regions, weighs nearly 20 kilograms (44 pounds) and is capable of producing only two chicks in a three-year period.

Graham Robertson

A group of emperor penguins crosses an ice field in characteristic single file.

◀ Most penguins walk on land, but the rockhopper penguin (far left) often bounds along with feet together, like someone in a sack race. Second-largest of penguins, the king penguin (near left) breeds in large dense colonies, some of which exceed 100,000 pairs. Severely persecuted by humans last century, the king penguin is now increasing rapidly in number. Both species inhabit subantarctic islands.

The chicks spend the winter in large groups known as crèches where they are fed sporadically, and many perish. The chicks require nine to thirteen months to fledge, the longest fledging period of any bird. Formerly exploited for their oil, most king penguin colonies have recovered since being given legal protection.

The Adelie, gentoo, and chinstrap penguins are collectively referred to as the long-tailed penguins. The most familiar of penguins, the Adelie penguin *Pygoscelis adeliae* is essentially restricted to the Antarctic, where a minimum of 2½ million pairs breed. The chinstrap penguin *P. antarctica* occurs in an area known as the Scotia Arc, extending from the tip of the Antarctic Peninsula and including the South Shetland, South Orkney, and South Sandwich islands, and South Georgia. The gentoo penguin *P. papua* inhabits mainly the subantarctic although some breed along the north coast of the Antarctic Peninsula. All three species nest during the southern spring and summer, October to February. The migratory Adelie penguin winters in the pack ice, whereas wintering chinstraps favor open water. Some gentoo penguin populations remain near their colonies year round, but others disperse widely.

The six species of thick-billed crested penguins (genus *Eudyptes*) are essentially circumpolar throughout the subantarctic, and adults are characterized by prominent orange or yellow crests. All species lay dissimilar-sized two-egg clutches, and although some (at least three species) may hatch out two chicks, none is capable of fledging both young. Several species have very restricted breeding ranges—for example, the royal penguin *E. schlegeli* at Macquarie Island, and the Snares Island crested penguin *E. robustus* only at Snares Island.

The four species of basically non-migratory temperate and tropical penguins within the genus *Spheniscus* span the greatest latitudinal range: the New World species extend from the Galapagos Islands south to the tip of South America and the Falkland Islands. The black-footed penguin *S. demersus* lives in South African and Namibian waters. Most spheniscids are burrow- or crevice-nesters, but at some crowded colonies they may nest on the surface.

The burrow-nesting little blue or fairy penguin *Eudyptula minor* of Australia and New Zealand is the smallest of the penguins—up to 30 would be needed to equal the weight of an emperor penguin. The yellow-eyed penguin *Megadyptes antipodes* of New Zealand is the most endangered penguin, with possibly fewer than 4,500 still extant, a decline of nearly 80 percent since the 1960s. Unlike most other penguins, nesting yellow-eyed penguins are not social, and if pairs are not visually isolated from one another they will fail to rear offspring. The loss of nesting habitat and the introduction of terrestrial predators have had a damaging impact.

FRANK S. TODD

DIVERS & GREBES

ORDER GAVIIFORMES
• 1 family • 1 genus • 5 species

SMALLEST & LARGEST

Red-throated diver *Gavia stellata*
Total length (extended): 53–70 cm
(20–27 in)
Weight: 1.1–1.7 kg (2½–3¾ lb)

White-billed diver *Gavia adamsii*
Total length (extended): 109 cm
(43 in)
Weight: up to 6.5 kg (14⅓ lb)

CONSERVATION WATCH
■ Diver numbers have declined,
but no species are listed as
threatened.

ORDER PODICIPEDIFORMES
• 1 family • 6 genera
• *c.* 19 species

SMALLEST & LARGEST

Least grebe *Tachybaptus dominicus*
Total length (extended): 24–28 cm
(9½–11 in)
Weight: 110–150 g (4–5 oz)

Great grebe *Podiceps major*
Total length (extended): up to
70 cm (27 in)
Weight: 1.5 kg (3½ lb)

CONSERVATION WATCH
!!! The Junín grebe *Podiceps
taczanowskii* and Alaotra grebe
Tachybaptus rufolavatus are
critically endangered.
!! The New Zealand dabchick
Poliocephalus rufopectus is listed as
endangered.
! The Madagascar grebe
Tachybaptus pelzenii is vulnerable.

Divers (known as loons in North America) and grebes (including the smaller dabchicks) are superbly streamlined aquatic birds about the size of a duck or small goose. Although classified in two separate orders—the Gaviiformes and Podicipediformes, respectively—they were once thought to be closely related, but the similarities are now considered to be the result of convergent evolution, involving independent specializations for diving. Genetic comparisons suggest that divers are related to penguins and petrels, which use their wings under water; and although today the divers are foot-propelled, some anatomical details suggest that they had wing-propelled ancestors. The affinities of grebes are still unclear, but certain anatomical features suggest relationships with sungrebes or rails.

DIVERS, OR LOONS

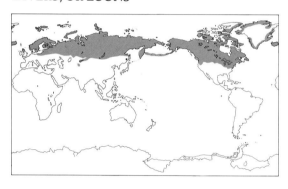

Divers are shy birds of lakes in northern forests and tundra. Their wild hollering and yodelling calls, audible for several kilometers, conjure up for many people the essence of these lonely northern wildernesses. After the breeding season the divers fly to their winter quarters at sea, where they molt their wing-feathers in October–December. They travel rapidly under water in pursuit of fish and in extreme cases can dive to a depth of 70 meters (230 feet). Although preferring rather shallow water, they usually stay well off the coasts, so the best chance to see them is when they migrate over certain projecting points along northern coasts.

In winter they are a rather uniform sooty gray above and white below, and the species are hard to tell apart, but for the breeding season their plumage is adorned with the most elaborate patterns. The red-throated diver *Gavia stellata* acquires its rufous throat-patch and a pattern of

▶ A great northern diver. Members of this family of waterbirds are known as divers in Europe, loons in North America.

Gary Neuchterlein

black and white lines on the hind-neck. The black-throated diver *G. arctica,* and its sibling species *G. pacifica* in eastern Siberia, Alaska, and Canada, have black and white lines on the sides of the neck and four white-checkered areas on their black back. The great northern diver *G. immer* of Iceland and North America, and the white-billed diver *G. adamsii* of Arctic North America and Siberia, are mostly black with their backs decorated by white square patches almost like a chessboard.

As they are unable to walk properly, divers place their nest at the lake's edge. This permits the bird to slip easily into the water. Most nest-sites are on peninsulas or small islands, places with a good view yet exposed to flooding from waves. Nests are usually just shallow scrapes in the boggy ground, but some pairs assemble reeds and waterweeds. The female lays two eggs on average, which are incubated for a month. The young are clad with sooty brown down and have two successive downy plumages (like the young of penguins and petrels). They can swim and dive at once, but the young of larger species prefer to ride on their parents' backs at first. The red-throated diver breeds mainly in small remote ponds and flies to larger lakes or out to sea for food. The other species prefer lakes large enough to satisfy the family's needs until the young fledge at 10 to 12 weeks of age.

GREBES AND DABCHICKS

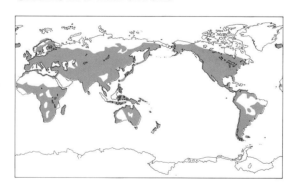

Slim, elegant, and adorned with crests and ruff, some grebes are considered the most fascinating birds on the lake. The best-known species is the great crested grebe *Podiceps cristatus,* widely distributed in temperate parts of Eurasia, Africa, and Australia. In North America its place is taken by the red-necked grebe *P. grisegena* and the western grebe *Aechmophorus occidentalis.* The smaller grebes, often called dabchicks, are less spectacular, resembling ducklings with small pointed bills. The feet, flexible and compressed, with semi-webbed toes, can work almost like a propeller, giving great maneuverability and speed when the birds pursue fish and invertebrates in the water.

▲ Grebes are notable for their spectacular courtship displays, which involve both sexes in a variety of highly ritualized movements. These western grebes skittering over the surface on rapidly pattering feet are performing a distinctive display known as "rushing".

Morten Strange/Flying Colours

▲ *A great crested grebe presents a feather to its chick. Feather-eating is common in several grebe species, but its purpose remains enigmatic: perhaps it is related in some way to a diet of fish.*

Grebes spend almost their entire lives in water. They even nest there, making floating platforms of plant material, which they usually conceal among reeds. The chicks can swim at once, but being sensitive to cold water they spend most of the time under the parents' wings.

Grebes have diverged considerably in their feeding adaptations. The primitive type of grebe is small and dumpy, living secluded among the reeds, taking water insects and making quick darts for small fish; such grebes exist worldwide, even in small creeks and ephemeral ponds. The large and streamlined grebes prefer larger lakes, for they are specialized for taking fish. The elegant western grebe can spear fish with its slender bill. Other dumpy and small-billed grebes are specialized to pick tiny arthropods from among the waterweeds. While fish-stalking grebes feed singly, the latter type are social feeders. Diet is largely determined by the anatomy of the bill, and where several closely related species occur in the same geographic area they have the most differently sized bills. Such "character displacement" reduces food competition between species.

Grebes usually avoid danger by diving, so flying is unnecessary for their daily life and they can therefore molt all wing-feathers at the same time. Some species congregate in very productive lakes when molting—750,000 black-necked grebes *Podiceps nigricollis* spend the fall at one lake in California. To become airborne the birds need a running start. They fly fast but, unable to make fast turns, may be vulnerable to birds of prey and therefore prefer to migrate at night. Northern

Hemisphere species spend the winter on icefree lakes or at sea. Grebes living under stable climatic conditions may not need to fly at all. A couple of species have entirely lost the power of flight and remain at one lake for their whole life: the Junin flightless grebe (puna grebe) *P. taczanowskii* in a highland lake in Peru, and the Titicaca flightless grebe (short-winged grebe) *Rollandia micropterum* in lakes of the Peruvian–Bolivan high plateau.

The grebes' courtship rituals are complex, and especially in the larger species the two partners often face each other for long bouts of mutual head-shaking. They also "dance", maintaining their bodies vertically almost out of the water, or they rush side by side in upright positions. The sexes look alike, but each sex is distinguished by fine voice differences.

THREATS TO SURVIVAL

The major threat to the survival of divers and grebes is the destruction of their freshwater habitat. Being specialized feeders, they are very sensitive to changes in their lake ecosystems—divers being adversely affected by acid rain, and grebes by pollution and draining of their shallow breeding habitat. Human activities also affect the breeding success of many species, especially that of the divers, who will readily surrender their nests if disturbed. Despite these threats, divers are not at risk of extinction, although numbers have declined in recent years. Two species of grebes, however, may already be extinct, and four species are listed in the IUCN list of threatened birds.

JON FJELDSÅ

PELICANS & THEIR ALLIES

One family in the order Pelecaniformes is particularly well known around the world—the pelicans of the family Pelecanidae. These ungainly comic characters with their large feet, waddling gait, long bill and huge pouch have attracted attention throughout recorded history. The other four families are tropicbirds (family Phaethonitidae), gannets and boobies (family Sulidae), cormorants and anhingas (family Phalacrocoracidae), and frigatebirds (family Frigatidae). The one characteristic that unites these diverse families is their totipalmate feet: all four toes are connected by a web. The most familiar feature of the group is probably the large naked throat (gular) sac, which can be spectacular in pelicans and male frigatebirds, but is entirely absent in the tropicbirds.

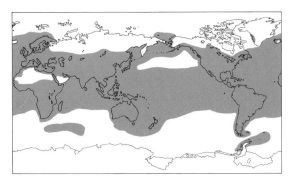

NESTING IN COLONIES

While primarily marine birds, pelecaniforms are found in all types of water environments from the open ocean and sea coasts to lakes, swamps, and rivers. Pelicans and cormorants may switch between fresh water and salt water during the year. For example, the American white pelican *Pelecanus erythrorhynchos* nests inland and migrates to coastal waters during the non-breeding season. Most species inhabit tropical and temperate regions, but several species of gannets and cormorants live in Antarctic and subantarctic waters.

Pelecaniforms commonly live more than 20 years, and many return faithfully to the same nest site each year, mating with the same individual. All species are colonial to some extent, nesting with members of their own type, in isolated places such as on islands. Cormorants and gannets often build

KEY FACTS

ORDER PELECANIFORMES
• 5 families • 7 genera
• 56 species

SMALLEST & LARGEST

White-tailed tropicbird *Phaethon lepturus*
Head–body length: 35–40 cm (14–16 in)
Central tail feathers: 35–40 cm (14–16 in)
Weight: about 460 g (1 lb)

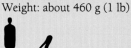

Dalmatian pelican *Pelecanus crispus* & Australian pelican *Pelecanus conspicillatus*
Total length: 160–180 cm (63–71 in)
Weight: *c.* 15 kg (33 lb)

CONSERVATION WATCH
!!! The Ascension frigatebird *Fregata aquila* is listed as critically endangered.
! 12 species are listed as vulnerable: Dalmatian pelican *Pelecanus crispus*; spot-billed pelican *Pelecanus philippensis*; Abbott's booby *Papasula abbotti*; Campbell Island cormorant *Phalacrocorax campbelli*; New Zealand king cormorant *Phalacrocorax carunculatus*; Stewart Island cormorant *Phalacrocorax chalconotus*; Auckland Islands cormorant *Phalacrocorax colensoi*; Pitt Island cormorant *Phalacrocorax featherstoni*; Galapagos cormorant *Phalacrocorax harrisi*; Chatham Islands cormorant *Phalacrocorax onslowi*; Bounty Islands cormorant *Phalacrocorax ranfurlyi*; Christmas Island frigatebird *Fregata andrewsi*.

Graham Robertson

◄ *A colony of Cape gannets in South Africa. Gannets nest in large dense colonies in which all stages of the breeding cycle are closely synchronized. The birds vigorously defend their territories, but these are so small as to be effectively bounded by the reach of the owner's bill while sitting on its nest.*

their nests within a few inches of each other. Aggressive interactions between neighbors are frequent in these dense colonies as each pair defends a space around their own nest. In contrast, masked booby colonies are often loose associations of birds nesting 6 to 60 meters (20 to 200 feet) apart. Frequently, a single colony will have two or three, or more, different pelecaniform species nesting in it.

In all species, both parents share incubation and chick-rearing duties. An individual's shift of sitting on the egg(s) varies from a few hours to more than a week, according to the nesting locale and the species. When the mate returns from feeding, the pair may go through a brief greeting display, allowing mate recognition and reinforcing the pair bond. If the incubating bird does not relinquish its duties, the returning bird, seemingly impatient to begin, may sit beside the bird on the nest and begin gently pushing it off. Incubation takes four to seven weeks depending on the species, and the adults either wrap the eggs in their large webbed feet or tuck them into the breast feathers to provide warmth. They do not have a brood patch. Chicks hatch naked in all pelecaniforms except the tropicbirds, whose single chick is covered by thick, fluffy grayish to white down when it hatches. Chicks must be carefully shaded from the hot sun during the day and kept warm at night. After two to three weeks, chicks are covered with down. By four to six weeks, they are often left alone while both parents forage for food. Adults regurgitate meals directly into the chicks' mouths—birdwatchers may wonder how any of the chicks get fed, when three nestling pelicans all have their long bills jammed into one parent's

▼ A white-tailed tropicbird soars over its breeding colony in the Seychelles. Except when breeding, tropicbirds are solitary inhabitants of the open ocean, distributed as their name implies. They are characterized by a pair of slender tail streamers equal in length to the rest of the body.

C.A. Henley

throat. Pelecaniform birds feed primarily on fish, but squid and other invertebrates and even other birds' eggs and small chicks may be eaten. In years when food is plentiful, more than one chick may be raised in species that lay more than one egg.

The air-sac system, which enables birds to have a continuous supply of oxygenated air passing through the lungs both when breathing out and when breathing in, is very extensive in many pelecaniforms. It also branches extensively across the chest and lower neck, making this area feel soft and cushioned to the touch. This padding may provide a shock absorber to species that dive into the water from the air (tropicbirds, pelicans, gannets, and boobies) and may provide extra buoyancy.

C.A. Henley

TROPICBIRDS

Tropicbirds differ in many characteristics and behaviors from other pelecaniforms, while sharing the defining character of the order: totipalmate feet, in which four toes (instead of three toes as in many other species) are connected by webs. They are widely distributed in tropical and subtropical seas, occurring in the Caribbean, Atlantic, Pacific, and Indian oceans, and are not seen on land during the non-breeding season. The three species (genus *Phaeton*) are similar in appearance, being white, with some black on the head, back, and primaries. They all have two long central tail feathers.

Amazingly, tropicbirds cannot walk—their legs are located too far back on the body—and to move on land they push forward with both feet and plop

on the belly. They perform their courtship rituals in the air, flying around in groups of two to twelve or more, squawking loudly. A pair will land and push their way into their potential nest site, a hole in a cliff or under a bush, where they sit and squawk at each other, somehow deciding whether or not to form a pair. Tropicbirds are often referred to as boatswain birds, presumably because their harsh call is reminiscent of a boatswain's yell. They are plunge divers, diving into the water from varying heights to catch their fish or squid prey.

PELICANS

All seven species of these fascinating birds have the characteristic pouch, long bill, and long neck. The adults of five species are primarily white, with

▲ *A group of Australian pelicans glides in for a landing. In terms of body weight, pelicans are among the largest of all flying birds.*

▶ The blue-footed booby, like other gannets and boobies, feeds on fish. The bird plunges in spectacular dives from a height of 2–15 meters (6½–50 feet), disappearing beneath the ocean surface in pursuit of its meal. The impetus of the dive takes it beyond the fish, which it snatches from below as it returns to the surface.

D. Parer and E. Parer Cook/Auscape International

▲ The brown pelican (left) is unusual among pelicans in that it plunge-dives for its prey in a manner similar to that of gannets and boobies. Second rarest of the world's boobies, the blue-footed booby (right) is restricted to the eastern Pacific Ocean.

some black in the primary and secondary feathers. The gray or spot-billed pelican *Pelecanus philippensis* is light gray with some black primaries. The brown pelican *P. occidentalis* has a very complicated sequence of annual adult plumage changes: it is silver gray on the back, with a black belly, white head, and white neck, but at different times of the year, depending on the breeding stage, the neck is chocolate brown and/or the head is yellow, and the pouch varies in color.

Most pelicans nest near and feed in fresh water, although all species are able to feed in either fresh or salt water. The brown pelican is the only marine species and is the only one to dive from varying heights into the water to catch a meal. The other species all feed as they sit on the surface, dipping down into the water with their bill and extensible pouch. Groups of pelicans will herd fish into shallow water where they are more easily caught. Brown pelicans especially are known to scavenge for meals around fishing piers and boats, unfortunately contributing to their decline in many places as they easily become entangled in fishing line which later snags on rocks or trees.

Adult pelicans are essentially voiceless, and courtship involves the use of "body language" (visual displays) rather than vocalization. Males pick a nest site in the colony—pink-backed *P. rufescens* and spot-billed pelicans mainly in trees; Australian *P. conspicillatus*, great white *P. onocrotalus*, Dalmatian *P. crispus*, and American white pelicans on the ground, and brown pelicans either in trees or on the ground—where they display by posturing, primarily with the head and neck, as females fly over. When a female is attracted, the two go through a series of coordinated displays. Once mated, the male brings nest material to the female who builds the nest. The number of birds in a colony ranges from as few as five pairs to several thousand. A large pelican colony is very noisy because chicks do have a voice and use it loudly to beg for food from their parents.

Leo Meier/Weldon Trannies

GANNETS AND BOOBIES

The three species of gannets live primarily in temperate regions, while the six species of boobies range throughout the tropical and subtropical regions of the world. Most boobies and gannets have a white head, neck, and underside, and are white with varying degrees of brown or black on the back. Colors of the soft parts of the bill and feet vary from bright sky-blue in the blue-footed booby *Sula nebouxii* to vivid red in the red-footed booby *S. sula*. Plumages are similar in both sexes, except the eastern Pacific subspecies of brown booby *S. leucogaster,* in which females have brown heads that match the body, and males have lighter brown to white heads. Male gannets are larger than females, but the reverse is true for boobies.

It is thought that the name "booby" is derived from the Spanish word *bobo* meaning clown or stupid fellow. Courtship can involve much parading around, lifting of the head up high, mutual preening, fencing with bills, and tossing of heads. When seen, it does make the booby's name seem appropriate. Only Abbott's *Papasula abbotti* and red-footed boobies nest in trees, building a nest of twigs which may be lined with some leafy vegetation. The others lay their egg(s) on bare ground or build a nest of twigs, debris, or dirt.

Gannets and boobies dive like missiles, often from amazing heights, into the water to capture fish and squid, feeding alone or in flocks. Sometimes gannets may even pursue fish underwater, moving with powerful feet and half-opened wings. A few boobies are reported to be kleptoparasitic, chasing other boobies until they regurgitate, then stealing the meal.

CORMORANTS AND ANHINGAS

Most of the 28 species of cormorants (also known as shags) and four species of anhinga (or darters) live in tropical and temperate areas, but some inhabit colder Antarctic and Arctic waters. Some species are solely freshwater, others solely marine, and some are found in both habitats. One member of the family, the Galapagos cormorant *Nannopterum harrisi,* cannot fly. It hops in and out of the water, scrambling up rock ledges to roost or get to its nest site on predator-free islands. Tree-nesting species construct nests of twigs, whereas species that nest on rock islands or cliff ledges use seaweed, or even guano and old bones, to build their nests. Cormorants form some of the largest and densest seabird colonies in the world which, as might be expected, produce great quantities of excreta. This guano is mined in some areas of the world for fertilizer.

Most cormorants and anhingas are black and may have an iridescent green or blue sheen, while others have striking white markings. Their diet is mostly fish but includes smaller amounts of squid, crustaceans, frogs, tadpoles, and insect larvae. They generally pursue their food by swimming underwater. The legs and feet, placed far back on the body, may not make walking easy but they make great propellers. The fish-catching ability of

▲ *Cormorants lack any waterproofing agent in their plumage. This reduces the energy needed to remain underwater in pursuit of fish, but it also means that eventually the plumage becomes too waterlogged to continue. Cormorants therefore spend much of their time loafing at the water's edge while their plumage dries out, like these pied cormorants.*

Zefa/Wisniewski

and sit in groups with other males, all with the large red gular sac expanded like a huge red balloon. When a female flies over, the males begin bill-clattering, whinnying, and fluttering the wings. Females land and display with various males until a pair determines they are "compatible". The male then begins bringing nest material to the female, frequently stealing twigs from unwatchful neighbors. A single egg is laid.

Frigatebirds probably have the longest chick-rearing period of any bird. The young begin to fly at five to six months but return to the nest to be fed until they are up to a year old. Flying juveniles are often seen "playing" with sticks or other items around the colony. One will dip down, picking up a stick from the ground, and others will pursue it, agilely coming at the young bird from all directions trying to grab the stick. The bird drops the stick, which may be picked up by another juvenile, often before it hits the ground, and the chase is on again. They are no doubt learning an important skill for catching their own meals.

Walking is impractical for frigatebirds because they have very short legs and small feet, and the very long tail drags if they land on the ground. Some ornithologists have suggested that they do not sit on the water because their feathers are not well waterproofed. With their large wings (wingspan up to 2.5 meters, or 8 feet), small feet, and small body, they do have a very difficult time taking off from water, getting the wings at the correct angle for flapping and eventual flying. Frigatebirds therefore feed on the wing, grabbing fish or squid from near the surface of the water or catching flying fish. With their huge wing area and light weight, they have the lowest wingloading of any bird measured and can remain in the air for days. They also have incredible maneuverability and often chase and harrass other birds, particularly boobies, causing them to regurgitate and then stealing the meal.

▲ A male greater frigatebird with its bright red gular pouch inflated. The frigatebird's courtship display ranks among the most bizarre of all avian activities. Males perch in bushes on their breeding islands: as females fly overhead, each male inflates his pouch, clappers his bill, and waves his wings wildly about in an attempt to coax the female to land beside him, whereupon they begin courting.

cormorants has been exploited by humans since the sixth century AD, and a few fishermen in Asia still keep trained flocks of cormorants. A collar is tied around the neck of each bird to prevent it from swallowing fish; then with a long line attached, the fishermen let the birds dive and catch fish, pulling them back to the boat and taking the fish when they resurface.

FRIGATEBIRDS

The five species of frigatebird range widely over tropical oceans during the non-nesting season and nest on isolated islands. All five are very similar in size and appearance: adult males are all black, except for the Christmas frigatebird *Fregeta andrewsi* and the lesser frigatebird *F. ariel* which have some white on the ventral side. Females are the larger sex, and all have some white markings on the underside, except the Ascension frigatebird *F. aquila* which is dark. Males pick out a nest site

THREATS TO SURVIVAL

The severest threats to pelecaniforms are caused by humans: disturbance of nesting colonies, and destruction of nesting habitat. The taking of birds for food causes the loss of many individuals. Predators such as rats, cats, and pigs introduced to islands continue to destroy breeding colonies; ground-nesters are particularly susceptible. Organochlorine pesticides such as DDT, which cause pelicans and other fish-eating birds to lay thin-shelled eggs that crush during incubation, are legally restricted, but the organophosphate pesticides now used are also highly toxic to birds. Oil spills and other water pollution cause local mortality. Pelicans are the most endangered of the pelecaniforms, probably because they live in close proximity to humans. Safe nesting habitats and unpolluted food sources are critical to their survival.

E. A. SCHREIBER

HERONS & THEIR ALLIES

T he order Ciconiiformes includes the herons, storks, ibises, spoonbills, and New World vultures, whereas the order Phoenicopteriformes is made up of the flamingos. Some of the families in these orders, together with the cranes (family Gruidae), are collectively known as the large waders. The New World vultures were until recently included with the raptors (order Falconiformes), but genetic and morphological studies have revealed that they are more closely related to the storks.

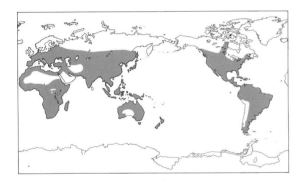

IDENTIFICATION BY BILL SHAPE

Large waders are so called because of their long legs and their generalized habit of wading in water to feed. (Small waders, also known as shorebirds, belong to the order Charadriiformes.) The feeding behavior and thus the bill shapes of the large waders are the best keys to their identification. The heron family shows the greatest evolutionary radiation with 61 species; they are recognized by their short straight bill, except for one or two species with unusually shaped bills for specialized feeding. Storks (17 species) are very large birds with correspondingly large bills of various shapes. Ibises (26 species) and spoonbills (6 species) belong to one family; ibises have sickle-shaped, downcurved bills, and the appropriately named spoonbills are easy to identify. The five species of flamingos have short, thick, curiously downcurved bills which immediately suggest an unusual feeding style.

After considerable debate about the taxonomic status of the New World vultures (family Cathartidae), genetic and morphological studies have justified their inclusion within the order Ciconiiformes. Before this, they were considered to be members of the order Falconiformes.

Large waders are found throughout the world, except near the poles, and in all habitats. Many species are gregarious, and frequently huge groups

KEY FACTS

ORDER CICONIIFORMES
• 7 families • 48 genera
• 117 species

SMALLEST & LARGEST

Little bittern *Ixobrychus minutus*
Total length: 27–36 cm (10½–14 in)
Weight: 46–85 g (1½–3 oz)

Andean condor *Vultur gryphus*
Total length: 1–1.3 m (3⅓–4⅕ ft)
Wingspan: up to 3.2 m (10½ ft)
Weight: 8–15 kg (18–33 lb)

CONSERVATION WATCH
!!! There are 6 critically endangered species: white-eared night-heron *Gorsachius magnificus*; dwarf olive ibis *Bostrychia bocagei*; northern bald ibis *Geronticus eremita*; crested ibis *Nipponia nippon*; black-faced spoonbill *Platalea minor*; giant ibis *Pseudibis gigantea*; California condor *Gymnogyps californianus*.
!! There are 7 species listed as endangered: white-bellied heron *Ardea insignis*; Australasian bittern *Botaurus poiciloptilus*; Chinese egret *Egretta eulophotes*; oriental stork *Ciconia boyciana*; Storm's stork *Ciconia stormi*; greater adjutant *Leptoptilos dubius*; white-shouldered ibis *Pseudibis davisoni*.
! 8 species are listed as vulnerable.

Belinda Wright

◄ *A little egret at its nest. Around the world, egrets were once slaughtered in vast numbers for their plumes. These specialized feathers, worn only during the breeding season, were known in the trade as "aigrettes" and were used to decorate women's hats.*

▲ *The painted stork (far left) is a large colorful Asian stork commonly seen in marshland reserves. The waldrapp or northern bald ibis (center left) is a cliff-nesting species of ibis from northern Africa. Highly endangered, it is the subject of an intensive conservation program. The green-backed heron (center right) is a common small species of the Americas, while the purple heron (far right) is a common large species of Europe and Asia.*

of several species can be seen feeding, roosting, or nesting together. When species are solitary they occupy ecological niches that seem to favor a lone existence. Most species are migratory and cover long distances to escape cold weather, probably because the weather readily affects the availability of their food and also because their large bodies consume a lot of energy to keep warm.

HERONS

Perhaps the most confusing thing about herons is their names; what are herons and what are egrets? The word egret is derived from the term "aigrette" which describes the filamentous breeding plumes found in six species of white heron. The definition has since been broadened by popular usage to cover several other heron species of other colors and lacking the fancy plumes.

Herons are among birds that have specialized feathers, called powder down, which are never molted but fray from the tip and continually grow from the base. The tips fray into a fine powder, which the birds use to remove slime and oil from their feathers.

Within the family Ardeidae there are three main types: typical herons, which generally are familiar

to most people; night herons, which are mostly active at night; and bitterns, which are rarely seen but often heard producing their characteristic "booming" call.

Typical herons

Typical herons are a diverse group varying in size, color, plumage pattern, and feeding behavior. Possibly the best-known species is the cattle egret *Bubulcus ibis* which has successfully invaded most areas of the globe. This heron has specialized its feeding behavior to benefit from catching insects disturbed by the grazing of large animals. This habit must have developed in association with wild animals such as buffalo, rhino, hippopotamus, giraffe, and elephant, but nowadays their link with domestic stock such as cattle and horses and their habit of following working plows have assisted their near-cosmopolitan spread.

Most typical herons are simple variations of one theme. They vary in size from the rufous heron *Ardeola rufiventris* to the giant Goliath heron *Ardea goliath;* they are monochrome (white, black, gray) or a mix of two or more colors including rufous, green, maroon, gray, and a huge array of shades; and some feed by standing still and waiting for prey to come to them, while others actively pursue prey by running or by flying and pouncing onto prey on the surface of water.

There are, however, some bizarre forms, such as the boat-billed heron *Cochlearius cochlearius,* which ranges from Mexico to Argentina. The bill of this bird has been described as slipper-like; it is much broader and deeper than the usual heron-bill and is flat above and curved below. The shape is accentuated when the bird displays its backward-drooping fan-shaped crest. The bill seems to be an adaptation to increase the area and number of prey sensors and thus the chances of snapping shut onto prey when the bird feeds at night. Some have suggested that the boat-billed heron is a weird variant of the black-crowned night heron.

The black heron *Egretta ardesiaca* of southern Africa has a remarkable feeding method. It inhabits shallow saline and fresh waters, where it bends its body forward, raises its wings above its back, and extends the wing tips downward to create an umbrella, casting a shadow on the water's surface. The bird's head is below this canopy of wings, which seems to act as a false refuge under which aquatic prey are tricked into sheltering.

Night herons
The night herons are so named because they feed mostly at night. They are short dumpy herons with rather thick bills and comparatively short legs. One

◄ The yellow-billed stork is common and widespread over much of Africa. A desultory feeder and only mildly gregarious, it wades in shallow water catching frogs and other small aquatic animals.

species, the black-crowned night heron *Nycticorax nycticorax,* is perhaps the most common heron in the world. They take a wide range of food items, including fish, frogs, snakes, small mammals, spiders, crustaceans, and insects. During the day most night herons roost quietly, usually high in trees, where the astute observer may see them.

Bitterns

Bitterns are inconspicuous, quiet creatures which inhabit densely vegetated wetlands and hence are rarely seen. They adopt cryptic postures, including standing fully erect with the bill pointed skyward, thus resembling the reeds and other thin vertical vegetation of their homes. They will even wobble slowly to resemble the movement of vegetation blown by the wind. They have a curious adaptation whereby the eyes are placed widely on the head which allows them to see across a wide field of vision even when they are sky-pointing. All of this reflects their dependence on camouflage rather than flight to avoid predators.

STORKS

Storks are best known through tales about just one of the 17 species. The white stork *Ciconia ciconia* of Eurasia is prominent in folklore because of its image of natality. Its habit of nesting on buildings, often on chimneys, and its return to the breeding grounds in spring after wintering in Africa have long been associated with human birth. But inspiring as this species is, other members of the stork family are also intriguing. For example, one unusual variation is shown by the shoebill *Balaeniceps rex.* Imagine a huge dark gray bird that seems to be wearing a Dutch clog-shoe on its face! Well, it's true; this bird is found in wetlands in Africa. Unfortunately, despite its immense appeal, it has not been studied in any detail. The two species of openbill storks (genus *Anastomas*) are also easy to recognize. Their huge bills seem to have been bent about half-way along when they were biting on something hard. In fact they use this gap to grasp and crush freshwater snails, which seem a strange food supply for such a large animal.

Storks as a group have been more intensively studied than most of the other large waders. This can be attributed to a single researcher, Phillip Kahl, who spent more than a decade traveling the world to study all 17 species. Kahl's main interest was to try to tease out the evolutionary relationships between the species. This could be achieved with storks because they have elaborate and complicated courtship displays which they perform conspicuously. The results of Kahl's work have been presented in a fascinating series of eminently readable scientific papers. Perhaps more importantly, Kahl's other passion is photography, and through an extensive series of photo articles published by his main sponsor, the National Geographic Society, general awareness of the need to protect wetlands and wetland birds has been heightened.

KIM W. LOWE

NEW WORLD VULTURES

Seven species belong to the family Carthatidae. This group from the Americas looks remarkably like the Old World Vultures of Africa, Asia, and Europe, which are true birds of prey. The New World vultures are distinguished by their perforate nostrils (open from one side of the head through to the other) and their lack of a functional hind toe.

All cathartids have rather dull plumage but an often colorful bare head and neck in adulthood. Corrugated, wattled skin in brilliant red, orange, and yellow adorns the head of the king vulture

▼ *The extraordinary shoebill is restricted to the interior of Africa, especially the Sudd region of southern Sudan. It resembles a heron rather than a stork in much of its behavior and, like herons, retracts its head back between its shoulders in flight.*

David C. Houston/Bruce Coleman Ltd

Francois Gohier / Auscape International

Sarcorhamphus papa. Inflatable neck pouches highlight the pink-yellow bare skin during courtship displays by the California condor *Gymnogyps californianus,* and the red skin on the face and neck of the turkey vulture *Cathartes aura* becomes brighter during conflict. The Andean condor *Vultur gryphus* has a thick white ruff of downy feathers around its neck. The sexes are similar in all species except the Andean condor, in which the male has brighter skin, and a comb on his forehead.

All New World vultures are scavengers; turkey vultures also take a few live prey such as young birds and turtles. The three members of the genus *Cathartes* have a keen sense of smell, which they use to locate carcasses, rejecting those that are decayed. The king vulture is often associated with the smaller turkey vulture or greater yellow-headed vulture *Cathartes melambrotus,* and follows them to carcasses, where it uses its strong bill to break open the body so that all can feed. All species can go for weeks without food, then gorge when they find it.

The turkey vulture occupies a wide range of habitats, from desert through grassland to dense tropical forest and temperate woodland. Recently, it has expanded its range, and is most common in disturbed areas. The American black vulture *Coragys*

atratus is even more closely associated with humans, and congregates in vast numbers around markets, garbage dumps, and fish wharves. In contrast, the huge Andean condor prefers the high mountains but ventures to sea level to feed on beach-washed whales, seals, and seabirds. The condor is becoming increasingly rare, largely because of persecution for its alleged attacks on livestock.

Cathartids do not build a nest, but lay one or two eggs on the ground or on a cliff ledge, or in a tree stump or cave. Both parents care for the young and regurgitate food from their large crops to feed them, unlike the true birds of prey, which carry prey to the nest in their talons. Populations in more northern areas congregate to migrate southwards for the winter; elsewhere they are resident, but may range widely in search of food.

PENNY OLSEN

IBISES

Ibises are another successful group that has radiated into most habitats and can be recognized by most people. They are strongly linked with wetlands, where they probe at and into soft sediments or water for their food. They also nest in wetlands, often in huge tightly packed colonies. However,

▲ A group of Andean condors congregates at a carcass. These are even-tempered birds, and squabbling while feeding is rare. Weighing up to 14 kilograms (31 pounds) and with a wingspread of 3 meters (10 feet), the Andean condor is the largest of the New World vultures.

71

▶ Spoonbills are found on all continents except Antarctica. All six species are very similar in general appearance, and all but one (the roseate spoonbill) have white plumage like this African spoonbill.

Jonathon Scott/Planet Earth Pictures

several species have adopted a different lifestyle. One species, the hadadah *Bostrychia hagedash* of central and southern Africa, utilizes wet and dry forests for feeding and breeding. The bald ibis *Geronticus calvus* is confined to southern Africa in an area where lightning strikes have burned much of the native forest and large trees are rare; this adaptable ibis now nests on sandstone cliffs, often beside waterfalls that are inaccessible to predators.

Unfortunately, ibises have more endangered species than the other families of large waders. The waldrapp *G. eremita,* a close relative of the bald ibis, is now restricted to a small population at only a few sites. The Japanese crested ibis *Nipponia nippon* was once widespread in Japan, but now only a handful of birds survives there in the wild, with another small population in China.

SPOONBILLS
These unusual birds are well liked throughout the world. The elongated and flattened bill resembles a spoon, which is whipped (partly opened) from side-to-side in sweeping strokes through the water until a prey item is touched. Acutely sensitive nerve endings lining the bill register the vibrations caused by the prey as it is touched, and the bill snaps shut in an instant. The bird then throws its head back, releasing its grip on the prey which falls down into the throat. Watching spoonbills feed has caused me to wonder why they don't get

headaches or feel ill from these jerking movements of their heads. Intensive studies of spoonbills are few, but work in Australia on the royal spoonbill *Platalea regia* has shown that these birds can be voracious feeders, sometimes consuming hundreds of prey items per day. In some marine ecosystems where they concentrate on a shrimp diet, their selective feeding on large, mainly female, shrimps seems to change the sex ratio in the shrimp population, so the number of shrimps that the spoonbills consume has an important effect on the whole ecosystem. This species can be found alongside the yellow-billed spoonbill *P. flavipes* at inland sites in Australia, where contrast between the birds and their feeding behavior is strong: the royal has a shorter, fatter body and bill than the yellow-billed; the former's bill is moved more rapidly through the water as the bird wades forward. These differences are reflected in the type of prey that the species catch at the same place: the royal catches the faster-moving prey such as fish.

Apart from the European spoonbill *P. leucorodia*, other spoonbills are not so well known. The beautiful African spoonbill *P. alba* is poorly named because all spoonbills have predominantly white plumage; rather it should have been named after its pink legs, bill, and face. The black-faced spoonbill *P. minor* inhabits the Orient but is not commonly reported. The roseate spoonbill *P. ajaja,* which inhabits the coastal swamps of the Americas, is well named and presents an exciting splash of color in its drably colored habitats.

FLAMINGOS

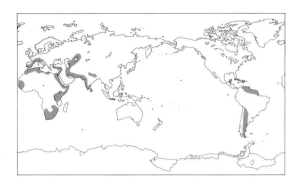

The bright pink of the flamingo's feathers and the strongly hooked bill instantly identify these birds. At first sight the birds look strange and awkward because their necks and legs are longer in proportion to their bodies than in other birds, but their beauty and grace are soon revealed.

KEY FACTS

ORDER PHOENICOPTERIFORMES
• 1 family • 1 genus • 5 species

SMALLEST & LARGEST

Lesser flamingo *Phoenicopterus minor*
Total length: 100 cm (40 in)
Weight: 1.9 kg (4⅕ lb)

Greater flamingo *Phoenicopterus ruber*
Total length: 140 cm (55 in)
Weight: 2.1–4.1 kg (4⅖–9 lb)

CONSERVATION WATCH
! The Andean flamingo *Phoenicopterus andinus* and Puna flamingo *Phoenicopterus jamesi* are listed as vulnerable.

THE SACRED IBIS & THE WISDOM OF THOTH

The sacred ibis *Threskiornis aethiopicus* had great importance in the everyday life of the ancient Egyptians. To these people certain animals symbolized individual gods, and the animals were worshiped as divinities. The sacred ibis was a symbol of Thoth, the god of writing and wisdom. This marvelous papyrus painting depicts the ibis-headed Thoth at the after-life ceremony involving weighing the heart of a deceased person. If the heart outweighed the feather of truth and judgment, it would be devoured by another animal-god; the result was recorded by Thoth, who reported it to the assessor-gods. Live birds were kept in temples and were mummified after death—huge collections, sometimes numbering half a million mummified birds, have been found in animal necropolises near the wetland breeding sites. Possibly, many of these were nestling birds that died of natural causes and were collected by the animal-protectors as a sign of respect.

Unfortunately, the ibises' importance in human culture did not stop this species becoming extinct in Egypt by the mid-nineteenth century, although it still thrives elsewhere.

▶ *This Egyptian papyrus painting dating from about 1250 BC depicts Thoth, the ibis-headed god of writing and wisdom, recording the result of the heart-weighing ceremony.*

C.M. Dixon

Frans Lanting/Minden Pictures

▲ A group of lesser flamingos feeding in the shallow waters of Lake Nakuru, East Africa. Lesser flamingos are more commonly found feeding by night in waters so deep that the birds swim rather than wade.

Flamingos are or were found on every continent, although only fossils are present now in Australasia. They seem to prefer salty or brackish waters through which they drag their bills upside-down. The upper mandible is lined with rows of slits, and the tongue is covered with fine tooth-like projections. The bill is opened, and then as the lower mandible is closed, water and mud are pumped out through the bill-slits. The residue probably contains a mix of microscopic food, which is swallowed.

The best known species is the greater flamingo *Phoenicopterus ruber* of Eurasia, Africa, and Central and southern America; it occasionally wanders from the Caribbean islands to Florida in the USA. All flamingos nest on lakes in shallow mud-cups scraped together so that they project above the water level. Only one egg is laid, and it takes about 30 days to hatch. The young birds leave the nests and herd together in large crèches, and are able to run and swim well at an early age.

KIM W. LOWE

RAPTORS

T he Latin word raptor means "one who seizes and carries away", a term that immediately calls to mind the archetypal eagles, the most powerful of avian predators. Yet the order Falconiformes—one of the largest and most fascinating of all avian groups—includes an unexpected range of form and habit, from swift, bird-catching falcons, among the fastest of birds, to huge, ugly, carrion-eating vultures. Their great strength and remarkable powers of flight and sight have inspired, enthralled, and terrified humans for thousands of years. So potent is their image that even today they appear on the crests of many nations, as military insignia and the logos of many businesses, and terms such as "eagle-eyed" have universal meaning.

KEY FACTS

ORDER FALCONIFORMES
• 3 families • 76 genera,
• 300 species

SMALLEST & LARGEST

Black-thighed falconet
Microhierax fringillarius
Total length: 14–17 cm (5½–6½ in)
Wingspan: 30–34 cm (11½–13⅓ in)
Weight: 28–55 g (1–2 oz)

Himalayan griffon *Gyps himalayensis*
Total length: 116–150 cm (45–58½ in)
Wingspan: 260–310 cm (101–120 in)
Weight: 8–12 kg (18–26½ lb)

CONSERVATION WATCH
!!! There are 3 critically endangered species: Madagascar serpent-eagle *Eutriorchis astur*; Madagascar fish eagle *Haliaeetus vociferoides*; Philippine eagle *Pithecophaga jefferyi*.
!! There are 8 species listed as endangered: Gundlach's hawk *Accipiter gundlachi*; imitator sparrowhawk *Accipiter imitator*; Hispaniolan hawk *Buteo ridgwayi*; red goshawk *Erythrotriorchis radiatus*; gray-backed hawk *Leucopternis occidentalis*; Javan hawk eagle *Spizaetus bartelsi*; Mauritius kestrel *Falco punctatus*; plumbeous forest falcon *Micrastur plumbeus*.
! 19 species are listed as vulnerable.

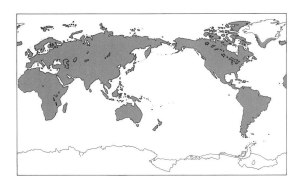

HOOKED BILLS AND SHARP TALONS

Raptors are characterized by hooked bills, strong feet, sharply curved talons, and large eyes. Their nostrils are surrounded by a fleshy cere straddling the base of the bill. Typically, they hunt by day—which is why they are often called diurnal birds of prey—and eat live prey caught with their feet. Also characteristic is the crop (a pouchlike extension of the gullet) to store freshly eaten food.

Differences in the basic design are adaptations to the extraordinary variety of raptor behavior. The vultures soar on long, broad wings and have strong feet and relatively straight talons; they must search vast distances to find carrion, then hold it while they tear with their bills rather than subdue it. In general, the falcons have a muscular body, long, pointed wings and long toes—all features necessary for swift, agile flight and capture of airborne prey. Many of the kites have a relatively slim body and weak fleshy feet, reflecting their generally less predatory habits and their scavenging or collecting of easily caught prey. Long-legged harriers and sparrowhawks reach into bushes or grass; sturdy-legged falcons hit prey in the air, often with great force. Some of the large forest eagles have massive legs and talons, and capture monkeys, sloths, and other large mammals from the trees. Falcons kill their prey with a blow or by biting the prey's neck and severing the spinal chord; hawks and other members of the family Accipitridae kill by the force of their grip, often compulsively squeezing the victim. All raptors use their feet to hold prey and their bills to dissect it.

The plumage of most birds of prey is brown to chestnut, dark gray, or black, often with mottled, barred, or streaked undersides. The Australian race of the variable goshawk *Accipiter novaehollandiae* can be pure white. A few species are adorned with crests, which they can raise in emotion. Some species have a distinctive juvenile plumage, and in a few the adult male differs in color from the female. Females are almost always larger than males, particularly in the agile bird-catching species; the diminutive male Eurasian sparrowhawk *A. nisus* is about half the weight of the female. Raptors that depend on immobile prey are the least dimorphic; the male of these species is about the same size as the female.

Raptors are renowned for their powers of flight and sight. From a distance of 1.5 kilometers (1 mile) a wedge-tailed eagle *Aquila audax,* a similar species to the golden eagle *A. chrysaetos,* can distinguish a rabbit from its surroundings; a human must approach to within 500 meters (550 yards). All raptors locate their prey primarily by sight.

Not least because they are prominent predators, raptors have suffered from pesticide contamination, habitat loss, and persecution, and some have long been coveted for falconry. Yet their image and appeal has also been to their advantage: considerable effort has gone into conserving species.

HABITATS

Raptors can be found in almost any habitat: from Arctic tundra to equatorial rainforest; arid desert to damp marshland; farmland to city. Because structural features of the habitat, rather than plant type, are most important to raptors, woodlands around the world tend to support a similar range of species. And raptors are not evenly distributed around the world: more than 100 species breed in the tropics, but only four in the high Arctic. Some habitats support raptors only at certain times of the year. For example, the rough-legged hawk *Buteo lagopus* of North America and Eurasia breeds above the treeline on open Arctic tundra, and then the entire population moves south to winter in farmland and marshes. Eleonora's falcon *Falco eleonorae* deserts its barren, rocky nesting islands

▶ National symbol of the United States of America, the bald eagle is now much reduced in number, and remains common only in Alaska.

Tom & Pat Leeson

in the Mediterranean and coastal Morocco in winter and migrates to the humid forests of Madagascar.

Some raptors can occupy a wide range of habitats as long as there is suitable prey to catch. At the other extreme are the highly specialized species, dependent on a particular type of habitat. The snail kite *Rhostrhamus sociabilis* eats only snails, collecting them in the freshwater lowland marshes of Florida, Cuba, and Mexico, south to Argentina.

Isolated islands often support lone species. The Galapagos hawk *Buteo galapagoensis* is the sole raptor on the semi-arid Galapagos Islands. In New Zealand there are two diurnal raptors: the New Zealand falcon *Falco novaezeelandiae* in the forests and the swamp harrier *Circus approximans* in the open grasslands. The raptor with the most limited distribution is the endangered Hawaiian hawk *Buteo solitarius*, confined to one island.

Habitat and form are linked. Forest-dwellers such as goshawks tend to have short rounded wings for buoyant flight among the trees; open-country species have either long pointed wings for rapid flight (falcons), or long broad wings for effortless soaring (buzzards and eagles).

THE RANGE OF RAPTOR TYPES

The raptors alive today can conveniently be divided into 13 general types.

Family Accipitridae

Osprey The osprey *Pandion haliaetus* is cosmopolitan in distribution, although in the Southern Hemisphere it breeds only in Australia. Its nostrils close as it makes spectacular dives into water for fish, sometimes submerging completely, either into the sea or into lakes. A reversible outer toe, long curved talons, and rough spiny toes help to grasp the slippery prey. Pesticides and persecution have decimated some populations, but local recovery programs have had good success.

Honey buzzards and white-tailed kites These birds are quite varied in appearance and habits. Some are extremely specialized: honey buzzards (especially the genus *Pernis*) eat the grubs and nests of wasps; the bat hawk *Machaerhamphus alcinus,* which occurs in Africa and from Burma to New Guinea, catches bats at dusk; the hook-billed kite *Chondrohierax uncinatus,* of Mexico south to Argentina, neatly extracts land snails from their shells with its deeply curved bill; bazas (genus *Aviceda*) eat insects, much preferring green grasshoppers to brown, and occasionally figs. The small gray and white kites (genus *Elanus*), with an almost cosmopolitan distribution, are rodent-eaters and accomplished hoverers.

True kites and fish eagles The black kite *Milvus migrans* eats many different foods but, with its taste for carrion, excreta, and other refuse, can mass in hundreds around abattoirs and villages. Its large cousins, the sea eagles *Ichthyophaga* and *Haliaeetus* (including the bald eagle *H. leucocephalus)* snatch live fish from near the water's surface, catch small mammals, and scavenge. They are also pirates, stealing prey from other birds, even other raptors.

Old World vultures Vultures are scavengers that rarely kill prey. They are incapable of sustained flapping flight and depend on rising air currents to keep them soaring aloft. Because of the vast distances they can travel in search of carcasses, they are exceptionally efficient and important scavengers. Most species have bare areas of skin on their head and neck, thought to reduce fouling of feathers when feeding and perhaps to help with heat regulation. The seven species of griffon vultures (genus *Gyps)* are usually numerous, and several hundred may gather at a food supply. They spread themselves in the sky and keep an eye on their neighbors; when one

bird spots a carcass and descends, many others follow. They gorge on muscle and viscera deep within a beast, and their ample crops can hold about one-quarter of the bird's body weight. The Eurasian black vulture *Aegypius monachus* is usually dominant over other vultures bickering at a carcass; its powerful bill enables it to eat coarse tissue, tendon, and skin. The Egyptian vulture *Neophron percnopterus* cracks eggs and eats their contents; it picks up stones and hurls them at eggs too large to pick up and drop. Bones and tortoises are dropped from a height repeatedly by the bearded vulture *Gypaetus barbatus* which swoops down to eat the fragments; its habitat is

Jean-Paul Ferrero/Auscape International

◄ With its large size and heavy, powerful bill, the lappet-faced vulture usually dominates the various vulture species that quickly congregate at any carcass on the African plains.

▼ The white-bellied sea-eagle is an Australasian species. It is not confined to the coast, being common along major rivers and around larger lakes in the interior. It catches fish by spectacular plunges from a height above the water surface, but it also patrols shorelines for carrion.

Belinda Wright

▲ *The extraordinarily flexible legs of the African gymnogene can be bent up to 70° behind and 30° from side to side, enabling it to grope in awkward tree hollows for the nestling birds and similar small animals on which it feeds.*

▼ *As a group, the harriers are birds of open country. Most species nest on the ground, and the spotted harrier of Australia is the only harrier that nests in trees.*

open mountainous country from Spain to Central Asia. The unusual palmnut vulture *Gypohierax angolensis* of Africa is dependent on the fruit of the oil palm.

Snake eagles Powerful feet and short rough toes help the members of this group to grasp snakes and other reptiles. They resemble the Old World vultures in some characters and, like them, spend long hours soaring. Most live in Africa, but the short-toed eagle *Circaetus gallicus* occurs also in southern Europe and eastwards to India. Although placed with this group, the bateleur *Terathopius ecaudatus,* an all-but-tailless bird of Africa, eats mostly carrion.

Harriers A uniform group (genus *Circus*) of long-winged and long-tailed hawks. They characteristically fly low and slow over open country and drop on small reptiles, birds, and mammals. Cosmopolitan in distribution, the 13 species do not overlap geographically except in Australia, where two species occur, the spotted harrier *C. assimilis* and the swamp harrier. The spotted harrier breeds in trees like most of the raptors. The other species nest on the ground in long vegetation. Some are polygynous; one male will pair with up to six females. When not breeding, some species are quite gregarious and roost communally on the ground.

Sparrowhawks and goshawks The genus *Accipiter* is the largest of the raptor genera, with about 48 species in woodlands and forests worldwide. Typically round-winged, long-tailed and long-legged, the largest is the northern goshawk *A. gentilis* (up to 1.3 kilograms, or almost 3 pounds), a powerful predator of large birds and mammals up to the size of hares. The smallest is the little sparrowhawk *A. minullus* (less than 100 grams, or 3½ ounces), which feeds on insects and small birds.

Typical buzzards and their allies Buzzards are medium to large hawks found in many habitats on all continents except Antarctica and Australia. Most are brown or gray in color. The 25 species of the genus *Buteo* soar readily on long broad wings and expansive rounded tails and catch mammals and reptiles on the ground. Harris's hawks *Parabuteo unicinctus* live and breed in small groups: perhaps four birds will hunt cooperatively, and after a jackrabbit is located it will be flushed by one bird towards the others waiting in ambush.

Harpy eagles These are very large, powerful eagles with unfeathered legs. The most powerful is the harpy eagle *Harpia harpyja* of tropical South America, followed by the monkey-eating eagle *Pithecophaga jefferyi* of the Philippines. Unusual among raptors, the latter has blue eyes.

True or booted eagles Legs that are feathered to the top of their feet distinguish these true eagles (including the genera *Aquila, Hieraaetus, Spiziastur* and *Spizaetus*), from the other eagles. They are

Tom & Pam Gardner

classic eagles in form and, in general, are active predators of mammals and ground-dwelling birds.

Family Sagittariidae

Secretarybird The secretarybird *Sagittarius serpentarius* is the only living representative in the family and has no obvious close relatives. It may not belong with the raptors, but is placed with them because it is eagle-like in appearance and in some habits. Large (standing about 1 meter, or 3 feet, tall) and semi-terrestrial, it strides over the African gasslands. Prey are subdued with a kick from the long legs, short stout toes, and nail-like claws.

Family Falconidae

True falcons About 38 species of the genus *Falco* and seven falconets comprise the true falcons. The largest are powerful birds capable of swift flight, which often seize or kill their prey with a mid-air blow. Measured in a stoop, a peregrine falcon *F. peregrinus* reached 180 kilometers per hour (112 miles per hour); the fastest of all birds, it also has the widest natural distribution of any bird. The gyrfalcon *F. rusticollis*, of the open Arctic tundra, captures ptarmigan flushed from the ground. Smaller kestrels habitually hover in search of ground prey; some are gregarious and live and breed in colonies. Most falcons breed as solitary pairs. The pygmy falcons (genus *Polihierax*), one in Africa and one in Southeast Asia, capture small birds and large insects on the wing.

Caracaras and others The caracaras are odd falconids. One species, the red-throated caracara *Daptrius americanus*, eats wasps and their nests, plus forest fruits and seeds. The other eight are scavengers. Unlike the true falcons, caracaras build a nest. Also of Central and South America are the forest falcons (genus *Micrastur*), which are five little-known primitive falcons superficially resembling harriers and goshawks.

▲ Kestrels of one species or another occur on all continents except Antarctica; illustrated here is the North American representative, the American kestrel (top left). The crested caracara (lower left) is one of a group of 9 species confined to Central and South America. They are chicken-sized birds that spend much of their time on the ground, hunting small animals or carrion. The wings of the bateleur (right), like those of a modern sailplane or glider, are unusually long and narrow, enabling it to swiftly cover large amounts of territory in search of food. It spends most of the day on the wing. In most other raptors the tail functions as a rudder, but the bateleur's relatively very small tail means that it must bank into turns in a highly distinctive manner.

Mitch Reardon/Weldon Trannies

▲ *The secretarybird of Africa differs from most other raptors in that it hunts on the ground, not from the air. It stalks about in open country, attacking prey ranging from grasshoppers to snakes and small mammals with powerful blows from its bill, or by stamping on them with its feet.*

naumanii typically nests in colonies of about 20 pairs, even as many as a hundred pairs, on a cathedral in Europe, an old fortification in southern Asia, or, less often, a well-recessed cliff. The kestrels feed on insect swarms and, like other colonial species, hunt, roost, and breed together.

Most raptors are ostensibly monogamous. A few raptors, such as some of the harriers, are polygamous. The Galapagos hawk and Harris's hawk are sometimes polyandrous, with several males attending a female. Almost all raptors have courtship displays, often with mock aerial battles. The male bald eagle *Haliaeetus leucocephalus* commonly dives at the female flying below. In response, she rolls and raises her legs to him. Occasionally, she grasps his feet and the pair tumble spectacularly earthwards. Typically male raptors offer food to the females (courtship feeding). The accipitrids (family Accipitridae), the secretarybird, and caracaras build a nest; the falcons appropriate the nest of another species or use a hollow in a tree, or a cliff ledge.

Typically, there is a clear division of labor during breeding: the male hunts, and the female carries out most of the incubation and brooding. She tears up food and offers it to the nestlings for the first few weeks after hatching. Unlike most other raptors, the secretarybird and honey buzzards regurgitate food from their crop to feed their young nestlings; and like the vultures, they also share nest duties more than is typical— behavior that is presumably related to the ease with which they can gather food.

Small species tend to have larger clutches, and shorter incubation and nestling periods than large species do: a kestrel or falcon may lay four eggs, have an incubation period of four weeks, and the young are in the nest for another five weeks. In contrast, the Eurasian black vulture *Aegypius monachus* lays one egg, incubates it for eight weeks, and the nestling is in the nest for 17 weeks. A few species lay two eggs but raise only one young— the first-hatched chick kills the second. Smaller species mature at an earlier age. For example, a kestrel can breed at the end of its first year, whereas a vulture must wait until it is five or six years old.

At the end of the breeding season some species migrate. Most prefer to migrate over land, funneling along land bridges such as the isthmus of Panama. Between September and November each year a vast stream of raptors passes over Panama on the way from North to South America for the winter: two million birds, mostly of two species (Swainson's hawk *Buteo swainsonii* and broad-winged hawk *B. platypterus*), have been counted, although many more go undetected. They travel as far as 11,000 kilometers (7,000 miles), taking as long as two months. Other raptors remain in their breeding area year-round, and some are nomadic and wander to wherever prey is available.

PENNY OLSEN

TERRITORIES AND MIGRATION

Generally territorial, in suitable habitat raptor pairs space themselves fairly evenly—about 1 square kilometer (⅓ square mile) for each kestrel pair, and 100 square kilometers (39 square miles) for each golden eagle pair. For some species the territory is permanent, but for others it is occupied and defended only during the breeding season; the latter may build a nest wherever prey is temporarily abundant. If a territory has a good nest site it may be occupied by successive pairs of raptors for centuries. The lesser kestrel *Falco*

WATERFOWL & SCREAMERS

KEY FACTS

ORDER ANSERIFORMES
• 2 families • 51 genera
• *c.* 160 species

SMALLEST & LARGEST

Indian pygmy goose *Nettapus coromandelianus*
Total length: 30–37 cm (12–14½ in)
Weight: 185–310 g (6½–11 lb)

Trumpeter swan *Cygnus buccinator*
Total length: up to 150 cm (59 in)
Wingspan: up to 3 m (10 ft)
Weight: 9.4–11.9 kg (20–26 lb)

CONSERVATION WATCH
!!! The 4 critically endangered species are: Madagascar pochard *Aythya innotata*; Brazilian merganser *Mergus octosetaceus*; pink-headed duck *Rhodonessa caryophyllacea*; crested shelduck *Tadorna cristata*.
!! The Madagascar teal *Anas bernieri* and white-winged duck *Cairina scutulata* are listed as endangered.
! 19 species are listed as vulnerable, including: Hawaiian duck *Anas wyvilliana*; swan goose *Anser cygnoides*; ferruginous duck *Aythya nyroca*; red-breasted goose *Branta ruficollis*; blue duck *Hymenolaimus malacorhynchus*; marbled teal *Marmaronetta angustirostris*; freckled duck *Stictonetta naevosa*.

Collectively known as waterfowl or wildfowl, ducks, geese, and swans are among the most beautiful of all birds. They are included in the family Anatidae and thus are often referred to as the anatids. All species are remarkably similar in form. While the center of distribution is in the Northern Hemisphere, their distribution is worldwide except Antarctica. Among the first birds to be domesticated, ducks and geese have been raised as a food source for more than 4,500 years, and all domestic varieties have been derived from the mallard, muscovy, graylag, and swan goose. The three species of South American screamers are the closest surviving relatives of the anatids, although superficially screamers more closely resemble gallinaceous birds such as the domestic fowl.

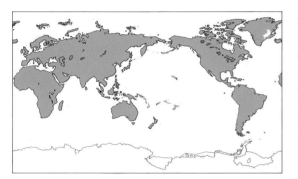

FLYING, SWIMMING, FORAGING
Most species are powerful fliers—although several are flightless—and many northern forms are highly migratory. Flying with continuous wing beats their speeds can exceed 110 kilometers per hour (68 miles per hour), but usually are considerably slower. Topographic barriers such as mountain ranges and oceans are not major factors in determining their distribution because most species can fly over them. Bar-headed geese *Anser*

Derrick Hamrick

◄ *The hooded merganser inhabits forested lakes and swamps in North America.*

81

indicus have been observed flying near the summit of Mount Everest at an altitude of 8,500 meters (28,000 feet).

With comparatively short legs and strongly webbed front toes, most waterfowl are excellent swimmers. Exceptions are the Hawaiian nene goose *Branta sandvicensis,* the Australian magpie goose *Anseranas semipalmata,* and the Cape Barren goose *Cereopsis novaehollandiae,* which have adapted to more terrestrial lifestyles and have less-webbed feet. The usual swimming speed is about 3 to 5 kilometers per hour (2 to 3 miles per hour), but faster speeds can be attained if the birds are pursued. Some species typically come ashore to roost, but others, such as the sea ducks and stifftails, generally doze on the water.

Most waterfowl have relatively long necks and flattened, broad, rather spatulate bills. However, the six species of mergansers differ, in that the bill is long, hard, and slender, with tooth-like serrations along the edges to aid in grasping slippery fish. In general, variations of bill shape reflect different feeding techniques; waterfowl feed on a wide variety of food items, and while grass, seeds, cultivated grain, and aquatic vegetation are favored by some, others will seek fish, mollusks, crustaceans, and insects. Aquatic invertebrates are particularly important for many ducklings and

goslings, even though as adults they may feed primarily on plants. There are three main methods of foraging: grazing, surface feeding, and diving. True geese are noted for grazing and spend a great deal of time foraging ashore. Surface feeding is the most typical, not only for the least specialized of wildfowl, but also for some of the most specialized such as the Australian pink-eared duck *Malacorhynchus membranaceus* which filter-feeds mainly on blue-green algae. Dabbling ducks are primarily surface feeders, typically ingesting material from the surface of the water, but they also upend in order to obtain food from the bottom. Diving ducks submerge for their food and rarely leave the security of the water because the rearward position of their legs makes it difficult to move on land. Most rarely dive deeper than 3 to 6 meters (10 to 20 feet) although there are exceptions. The oldsquaw or long-tailed duck *Clangula hyemalis* and the king eider *Somateria spectabilis* can dive to at least 55 meters (180 feet). Sea ducks feed chiefly in salt water and have evolved efficient salt glands, located above the eyes, to facilitate removal of excess salt.

All wildfowl, except for the fish-eating mergansers, have a highly functional, sensitive, fleshy tongue which is lined with many small spiny projections. The action of the tongue, working

▼ *A gallery of waterfowl. The northern screamer (below, left) is a member of a group of three species restricted to South America and isolated in the family Anhimidae. Other waterfowl belong to the cosmopolitan family Anatidae: top right, the red-breasted goose of western Asia; far right, the Carolina or wood duck of North America, and bottom right, the king eider, an inhabitant of the high Arctic.*

against the rows of horny lamellae that line the mandibles, functions as an efficient food-sifting mechanism. The peculiar sound emitted during feeding is known as "chattering", and except for the flamingos (which some ornithologists believe are closely related) no other group of birds feeds in this specialized manner.

All anatids are densely feathered with compact waterproof plumage, as well as a thick coat of insulating down beneath. Of all birds thus far examined, the tundra (whistling) swan *Cygnus colombianus* has the greatest number of feathers—more than 25,000—most of which are on the head and neck. While a number of species are dull and nondescript, many others are brightly colored and patterned. Males are generally more ornate and vividly colored, but with some birds such as the Siberian red-breasted goose *Branta ruficollis* both sexes are colorful. During the annual wing-molt the flight feathers are shed simultaneously, resulting in flightlessness for a period of three to four weeks. Only the Australian magpie goose (and possibly also the continental population of South American ruddy-headed geese *Chloephaga rubidiceps*) has a graduated wing-molt, so it avoids a vulnerable flightless period. The drakes of many northern ducks change their colorful nuptial plumage to a cryptic female-like "eclipse" plumage during the summer.

Waterfowl vocal abilities vary considerably in tone, intensity, and quality. Most people are familiar with the typical duck quack, but wildfowl also honk, hiss, trumpet, grunt, bark, squeak, cluck, and coo. In many cases the sexes have different voices, the female's generally being lower. No species is totally mute, not even the so-called mute swan. A number of species also produce mechanical sounds: ruddy duck *Oxyura jamaicensis* drakes, for example, have specialized throat sacs that can be inflated and beaten upon with the bill to create a distinct drumming sound.

COURTSHIP AND NESTING

Well known for their highly ritualized courtship behaviors, waterfowl are unlike most birds that require only cloacal contact for fertilization, and males have a distinct erectile penis. Copulation typically occurs on the surface of the water, but there are several exceptions. Most anatids are monogamous, and polygamous behavior is confined to only a few forms. The duration of the pair bond is variable: lifelong for swans and geese; almost nonexistent for the muscovy duck *Cairina moschata,* a polygamous species of the neotropics. Many nest in the open, while others seek the security of cavities. Nest construction tends to be rather rudimentary because waterfowl do not carry nesting material, but use only material that can be reached from the nest site. Some species such as the snow goose *Anser caerulescens* and common eider *Somateria mollissima* nest in colonies. The

S. Nielsen/Bruce Coleman Ltd

most unusual nester is the black-headed duck *Heteronetta atricapilla* of South America which lays its eggs in the nest of a host species—other ducks, or gulls, rails, ibises, herons, coots, and even snail kites. In all but a dozen species the female alone incubates; the period varies according to the species, for ducks about four weeks on average. Most waterfowl line their nests and cover the eggs with a thick layer of down plucked from the breast. Except for the Australian musk duck *Biziura lobata* and the magpie goose, which are fed by their parents, waterfowl young must feed themselves from the very beginning.

WATERFOWL
Magpie goose
The magpie goose, or pied goose, of northern Australia is the most atypical of all the wildfowl: a gradual wing molt, long toes only partially webbed, polygamous breeding, onshore copulating, and adult feeding of young are not typical waterfowl traits. Almost exclusively herbivorous, this striking bird has a specialized strongly-hooked bill that is used to dig out food from the hard-baked clay during the dry season.
Whistling ducks
The eight species of essentially tropical, long-legged whistling ducks are noted for shrill whistling calls, often uttered in flight. Gregarious

▲ *The white-faced whistling duck is widespread in Africa and in South America. Like most whistling ducks it is strongly gregarious, and spends most of its time in groups or flocks at the margins of swamps, lakes and rivers. The sexes are alike in appearance.*

C.A. Henley

▲ *The black swan is common in most wetland habitats across southern Australia, but it tends to avoid the northern tropics. Recent studies have revealed an unusual flexibility in its breeding habits. It may nest whenever conditions are appropriate, and sometimes raises several broods in succession. Birds are ready to breed at 18 months of age. Desertion, divorce, and various kinds of polygamy are common in young breeders, but pair bonds become increasingly stable as the partners mature.*

for the greater part of the year, most are sedentary, but those in temperate regions are at least partly migratory. Unlike most anatid species, pairs of whistling ducks preen each other and the pair bond is very strong, possibly lifelong. In some regions, the ducks are considered detrimental to crops. One of the most cosmopolitan of all birds, the fulvous whistling duck *Dendrocygna bicolor* lives in North and South America, Africa, and India, with no significant variation in size or color.

Swans

Largest and most majestic of the waterfowl, the eight species of swan are indigenous to every continent except Africa and Antarctica. The northern swans are noted for unbelievably loud voices, which is reflected in the descriptive vernacular names of trumpeter, whooper, and whistler. Most swans construct huge nests, and that of the trumpeter swam *Cygnus buccinator* of North America may be a floating structure. Both sexes care for the young, which are called cygnets. Several species carry their cygnets on their backs.

True geese

The 15 species of true geese are highly gregarious and are confined to the Northern Hemisphere, with most breeding in Arctic or subarctic latitudes.

Geese can be long-lived, and some captive birds have attained ages of nearly 50 years. Many, but not all, are somberly colored. The true geese are not sexually dimorphic, but the lesser snow goose *Anser caerulescens* has a distinct dark color morph known as the blue goose, which at one time was considered a separate species. The familiar Canada goose *Branta canadensis* has no less than 12 races ranging in size from scarcely larger than a large duck up to the size of a small swan.

Shelducks and sheldgeese

These birds live mainly in the Southern Hemisphere. They are all highly pugnacious, particularly during the breeding season, and many of the larger species have hard bony knobs at the bend of the wing which are used effectively in combat. Other than the northern swans, the adult kelp goose *Chloephaga hybrida* gander is the only all-white-plumaged waterfowl.

The Cape Barren goose of Australia has traditionally been regarded as an aberrant sheldgoose, but may represent a transitional link to the true geese of the Northern Hemisphere.

Steamer ducks

Often linked with the shelducks, all but one of these South American and Falkland Islands ducks

C.A. Henley

◀ Native to central Asia, the mute swan has been raised in a semi-domesticated state since the time of the Ancient Greeks. It is now common on ornamental waters across Europe, with a substantial feral population. In England, all mute swans by law belong to the Crown, a relationship extending back to the twelfth century. During recent decades, numbers declined in Britain because swans were poisoned when they ingested the lead weights lost or discarded by anglers. Use of lead weights was banned in 1987, and swan populations have partly recovered.

▼ The Canada goose originated in North America, but it has been widely introduced elsewhere and is now common in a feral state in, for example, New Zealand and Great Britain.

are flightless, but are capable of "steaming" over the surface of the water with flailing wings at speeds up to 18 to 28 kilometers per hour (11 to 18 miles per hour). Excellent divers, the aggressive steamer ducks feed primarily on shellfish and other marine organisms.

Perching ducks

This is a widely varying group, ranging in weight from a mere 230 grams to 10 kilograms (8 ounces to 22 pounds). As implied by their name, perching ducks are more arboreal than the other anatids and most are cavity nesters. Included in the group are the pygmy geese (*Nettapus* species) and the spur-winged goose *Plectropterus gambensis*. Drakes of the American wood duck *Aix sponsa* and Asian mandarin duck *A. galericulata* have such complex patterns and bright colors that they are considered to be among the most beautiful of all birds.

Dabbling ducks

In this, the largest group, which has some of the most familiar of all ducks, the genus *Anas* alone consists of 43 species including the mallard, wigeons, pintail, shovelers, and many teal. Also known as river or puddle ducks, the dabblers prefer freshwater habitats, but may frequent salt water, particularly during migration. An iridescent,

Stephen J. Krasemann/Bruce Coleman Ltd

▶ *Most widespread, versatile and numerous of the dabbling ducks, the mallard occurs naturally across much of the Northern Hemisphere, but it has been widely introduced elsewhere. This is of some concern to conservationists because the mallard hybridizes freely with local species in many parts of the world, compromising their genetic integrity. It is the ancestor of the domestic farmyard duck.*

Andy Purcell/Bruce Coleman Ltd

▼ *The common pochard belongs to a cosmopolitan group closely related to the dabbling ducks, but which dive for their food. Although flocks of some species often congregate in sheltered bays and inlets of the sea, on the whole they favor fresh water over salt water. This species is abundant across much of Europe and Asia.*

Richard T. Mills

metallic, mirror-like speculum on the wing secondaries is typical of both sexes of most *Anas* species.

Most dabblers breed during their first year and lay relatively large clutches of eggs. A new mate is courted each year. Best known is the mallard *A. platyrhynchos,* which ranks among the most successful of all avian species. No wild duck is more tolerant of human pressure and disturbance, and they have been able to adapt well to a rapidly-changing human-dominated world. Mallards and their near relatives are worldwide in distribution, and even occupy some remote islands.

Pochards

The pochards are rather closely related to the dabblers, but most are excellent divers. Unlike the dabblers, which can become airborne instantly from the water, the pochards (along with eiders and sea ducks) have to run over the surface of the water for a considerable distance before taking flight. Some species such as the greater scaup *Aythya marila* form up in huge rafts, principally at sea. The largest pochard, the canvasback *A. valisineria,* is the prime target for most North American duck-hunters because of its succulent taste. Sadly, the spread of agriculture over much of the prairie pothole country, as well as the drought of the late 1980s, has had a damaging impact on this magnificent duck and many other anatid species that breed in this habitat.

Eiders

The four eider species are specialized diving ducks, all of which inhabit the Arctic and subarctic. The famous eider down is thick and heavy, with the best thermal quality of any natural substance; in some regions such as Iceland, eiders are still "farmed" for their down. Drakes in nuptial plumage are among the fanciest of ducks, although the female plumage is a well-camouflaged brown. Three of the eider species are extremely heavy-bodied birds, but are strong fliers nonetheless. Much of their time is spent at sea, frequently in rough waters, eating aquatic vegetation and large quantities of mollusks, sea urchins, crabs, and other crustaceans.

Sea ducks

Scoters, goldeneyes, bufflehead, and mergansers are among the most accomplished of anatid divers.

With the exception of the rare Brazilian merganser *Mergus octosetaceus,* all sea ducks inhabit the Arctic and temperate regions in the Northern Hemisphere. While many do spend considerable time at sea, most nest inland near fresh water. The exquisite harlequin duck *Histrionicus histronicus* nests along fast-moving streams (often well inland), but gravitates to the sea during the winter. Its shrill whistling calls are clearly audible above the rushing roaring water. The oldsquaw is unique in that the drakes undergo three molts annually: the basically white plumage of winter is followed by an elegant brown with white for spring and early summer (it nests in the far north), and during the summer a dull-colored "eclipse" plumage is assumed. The largest merganser species, *Mergus merganser,* is known as the goosander in Britain because of its goose-like size.

Stifftails

Named for their distinctive stiff tails which the drakes often cock jauntily up into the air, the dumpy stifftails are so adapted to an aquatic, diving lifestyle that they can scarcely walk. During the breeding season the bills of the drakes of most species become a brilliant blue. Stifftails are essentially vegetarian, although the musk duck, a large Australian stifftail, is carnivorous. It is also the only stifftail that does not exhibit distinct sexual dimorphism in plumage; the male, however, is considerably larger than the female.

SCREAMERS

The three distinct species of neotropical screamers bear no superficial resemblance to typical waterfowl but are apparently closely related through a number of anatomical affinities. Like the magpie goose, they undergo a graduated wing-molt and thus do not have a flightless period. They are unique among living birds in lacking the uncinate process, the overlapping rib projection which strengthens the rib cage by serving as a cross-strut, a feature shared only with *Archaeopteryx,* one of the earliest-known avian fossils. In addition, screamers lack feather tracts; the only other birds that have feathers growing randomly are ratites, penguins, and African mousebirds.

Large, heavy-bodied birds with small heads and long legs, screamers are strong fliers and can soar for hours. Their very long toes exhibit only a trace of webbing, but despite this they are excellent swimmers although they do not dive. The long toes increase the surface area of the foot, distributing the weight and allowing them to walk over floating vegetation. The bill is distinctly fowl-like, and each wing is armed with two long, sharp, bony spurs. The largest species, the horned screamer *Anhina cornuta,* has a peculiar frontal "horn" that curves forward from the forehead toward the bill.

CONSERVATION

Despite the enormous pressures brought about by the increase in human population, most waterfowl species are holding their own. The long drought of the late 1980s and early 1990s severely reduced the numbers of North American species, but most have made a dramatic recovery. However, the loss of both breeding and wintering habitat continues to be a concern. Oil spills and other pollution also have a detrimental impact on waterfowl numbers.

FRANK S. TODD

▼ *In summer harlequin ducks inhabit rushing mountain streams in North America and Asia, but they spend the winter months at sea, congregating in bays, inlets, and fiords. The male is among the most striking of waterfowl, but the plumage of the female is mainly dark dingy brown.*

Tim Fitzharris

GAMEBIRDS & THE HOATZIN

The galliforms are familiar to almost everyone. "Chickens", which are domesticated forms of central Asia's jungle fowl, and turkeys from North America, are to be found on dinner plates around the world. Other gamebirds, such as pheasant, partridge, grouse, and quail, are known to hunters, and yet others such as the peafowl are noted for their beauty and ornate displays. Despite wide variation in size, all galliforms are characteristically stocky and have relatively small heads, plump bodies, and short broad wings. Flight is usually fast and low. Formerly considered to be related to the Galliformes, the strange-looking hoatzin is now placed in its own order.

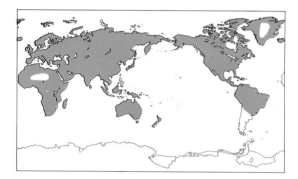

MEGAPODES

Members of the megapode ("large-footed") family have unique nesting habits. They lay eggs not in conventional nests, but in burrows or mounds. Thereafter parental care is restricted to maintaining the mound in such a way as to ensure stable temperatures for the eggs. The eggs are incubated by the sun, by organic decomposition of plant material or even by volcanic activity, and when the chicks hatch they are immediately able to fend for themselves. Of the 12 species distributed in Australasia and some Pacific islands, most inhabit rainforest and monsoonal scrub, but the best-known member of the family, the malleefowl *Leipoa ocellata* of southern Australia, inhabits semi-arid eucalypt woodland, where its numbers have declined markedly. All megapodes are vulnerable to human exploitation because the adults are conspicuous, largely ground-dwelling, tasty, and easy to shoot, and their eggs are large, highly nutritious, and easy to find. With widespread forest destruction, no fewer than eight species are endangered.

Several species build mounds as nests, the largest exceeding 11 meters (36 feet) in diameter and 5 meters (16 feet) in height. They are

▶ *Ready to lay, a female malleefowl approaches the mound while the guardian male looks on; he will drive her away if conditions are unsuitable for laying. Maintained entirely by the male, mounds are used year after year. Several females lay eggs in the mound at intervals throughout the nesting season, and the accumulated total may reach 30 or more.*

Tom & Pam Gardner

composed of leaf litter and soil and are used year after year. The malleefowl is strongly territorial, and the male maintains a mound of rotting vegetation throughout the year. Eggs are laid from September to January, in holes excavated in the organic matter by the male, who then re-covers the mound with sand and regulates its temperature at a constant 33°C (91°F) by changing the depth of sand as necessary. Other species on islands where soils are heated by volcanic activity lay their eggs in burrows where hot streams and gases provide the heat for incubation. Still more remarkable is the Niuafo'ou megapode *Megapodius pritchardii*, confined to Niuafo'ou island in the Tonga archipelago, which uses hot volcanic ash to incubate its eggs. The maleo *Macrocephalon maleo*, of Sulawesi in Indonesia, lays its eggs in tropical rainforest close to hot springs or on beaches exposed to the sun; egg collecting used to be supervised by rajas (kings) or other local authorities, and nesting grounds were leased and limits set for the number of eggs that could be taken. Unfortunately the loss of this traditional control has resulted in drastic overexploitation. Efforts are now being made to establish national parks and other protected areas where the maleo can survive alongside the local human population.

CHACHALACAS, GUANS, AND CURASSOWS

The Cracidae, or cracids, are found in the tropics and subtropics of the Americas, where they are largely arboreal (tree-dwelling). Within minutes of hatching the chicks show a remarkable instinct to climb and seek refuge in trees, and in a matter of days they can flutter from branch to branch in the canopy almost with the agility of adults. The smallest are the nine chachalaca species, which are the only ones not confined to humid forest, preferring the low woodland thickets more typical of the drier tropics. The guans (22 species) are medium-sized cracids and the most arboreal, while the curassows (13 species), which are the largest members of the family, spend up to half the day on the ground. In general the family is highly vocal, its members noted for the variety of their songs and cries echoing through the forest—the chachalaca name comes from its loud "cha-cha-la-ca" cry. In some species, notably guans, the windpipe is adapted for amplifying sound in the dense forest, resulting in the most far-reaching calls of all birds. Some species, in particular the curassows, perform elaborate courtship displays and courtship feeding; and many of the guans have a wing-whirring or drumming display (unique to the family), performed in flight.

TURKEYS

Two species of turkey exist: the common turkey *Meleagris galloparo,* native to the southern United States and Mexico; and the ocellated turkey *Agriocharis ocellata,* an endangered relative

confined to fragments of lowland rainforest in Mexico, Guatemala and Belize. It is believed that Mexican Indians were the first to domesticate the common turkey, which was subsequently introduced to Europe by the Spanish in the sixteenth century. Turkeys are large (males on average weigh about 8 kilograms, or 17 pounds; females 4 kilograms, or 8½ pounds), forest-dwelling birds with strong legs, large spurs and almost completely bare heads and necks. To attract females as mates, the males gather on strutting grounds known as leks. Here each male performs an elaborate display, spreading his tail fans, lowering and rattling his flight feathers and swelling up his head wattles. In this posture the males strut around, uttering their "gobble gobble" call. Because turkeys are polygynous (one male mating with several females), the dominant males achieve most of the successful matings at a lek.

▲ The wattled curassow is widely kept in zoos and similar collections, but little is known of its behavior in the wild. Curassows inhabit the rainforests of Central and South America. They feed mainly on the ground and roost in trees.

L.C. Marigo/Bruce Coleman Ltd

Mark Newman

A willow ptarmigan in its pure white winter plumage is superbly adapted to deep snow and bitter cold, right down to its feathered toes.

GROUSE

The more temperate zones of the Northern Hemisphere are the home of 17 species of grouse. Some live in coniferous forests, such as the spruce grouse *Dendragapus canadensis* and blue grouse *D. obscurus* of North America; others frequent more open habitats, such as the willow ptarmigan *Lagopus lagopus* of Europe's moorlands and mountainsides. In many northern areas, grouse are the largest year-round food source for predators, and their own numbers play an important part in determining the population size of some birds of prey. Some grouse undergo regular cyclic fluctuations in numbers, which have been studied extensively. Studies of the red grouse (a subspecies, in Britain, of *Lagopus lagopus*), for example, have shown that these cycles may be explained in some circumstances by fluctuations in parasite numbers. Other studies reveal variations in behavior of the grouse: aggressive birds hold large territories and so limit or decrease population growth; passive individuals tolerate crowding and so allow the population to rise again. Despite intensive investigations of blue grouse in North America and red grouse in Great Britain, the mechanism for such behavior is not yet fully understood. Humans also have an influence on gamebirds and their habitats: the desire to hunt game and therefore to maintain unnaturally high numbers leads people to control predators, provide additional food, and even release large numbers of artificially reared birds, which are all factors likely to distort natural population dynamics.

There is a wide range of reproductive social behavior. For example, the willow ptarmigan/red grouse is strictly territorial and forms a strong pair bond between male and female for the breeding season. In contrast, the sage grouse *Centrocercus urophasianus* males gather at large leks, each performing dazzling displays to attract females for a few moments, and the dominant male may well mate with more than 75 percent of the females in a "territory" not much bigger than two grouse.

NEW WORLD QUAILS

These are typically plump, rounded little quails, often boldly marked and with distinctive forward-pointing crests. All species studied so far are strictly monogamous and territorial during the breeding season; outside the breeding season they are gregarious, forming coveys or flocks. The subfamily includes three species of tree quail that inhabit montane forests in Mexico and Central America; the barred quail (*Dendrortyx* species) of Mexican arid scrub and woodland; the mountain quail *Oreortyx picta* of the woodlands of western USA; four grassland-dwelling North American species of crested quail, including the California quail *Lophortyx californica* (of which hunters shoot more than 2 million every year); four colins or bobwhites; and 14 species of wood quail

◄ Despite its name, the California quail is common and widespread in western North America from Vancouver Island to the tip of Baja California. It has been introduced to many other parts of the world. Both sexes wear the unusually-shaped topknot, but the female is otherwise much duller and browner than the male.

(*Odontophorus* species), medium-sized forest-adapted birds of Central and South America, two of which are threatened by deforestation in Colombia

OLD WORLD QUAILS AND PARTRIDGES

There are more than a hundred species in this group, occurring naturally from Africa east to New Zealand and north throughout most of Eurasia. The majority of species, however, are found in tropical Asia and in Africa south of the Sahara. A few have been introduced into North America.

Old World quail are small, rounded birds with short legs and relatively pointed wings. The common quail *Coturnix coturnix* is the only gamebird that regularly migrates from breeding grounds in Europe and central Asia to winter in Africa and India. Two other species, the harlequin quail *C. delegorguei* and the blue quail *Excalfactoria adansonii,* are nomadic and will move into areas in large numbers following rain; they inhabit grasslands and appear to show marked fluctuations in their population sizes.

The partridges are a much more diverse group: the very large, mountain-dwelling snowcocks which live in alpine zones from the Caucasus to Mongolia; the almost-unknown monal-partridges of China; seven species of red-legged partridges of Europe and the Middle East, adapted for life in arid regions; the larger francolins, predominantly of Africa; the spurfowl of Asia; the poorly known group of Southeast Asian species that inhabit tropical rainforests; and the so-called "typical" partridges, which were originally natural-grassland dwellers of northern Asia and Europe. The best-known is undoubtedly the gray partridge *Perdix perdix,* a native of Europe and parts of the Soviet Union but successfully introduced into North America. It has the largest clutch of any bird: 20 eggs. Throughout its range this bird is a prize

▶ The green peafowl is an even more handsome bird than the familiar blue peafowl, but it is somewhat more difficult to maintain in captivity, and hence not as widely known. It inhabits Southeast Asia.

C.A. Henley

▼ A male red junglefowl. This bird, the ancestor of the domestic chicken, was living in a semi-domesticated state in the Indus valley at least 5,000 years ago. It has since been taken by humans to virtually every corner of the globe, and it may well now be the most numerous bird on earth.

quarry for hunters, and although it initially adapted well to the intensification of agriculture in Europe it has since suffered an estimated 90 percent decline in population there, especially because of pesticides.

Almost all of the quails and partridges are monogamous, and pairs with their immediate offspring form the basic social unit, a covey. In the relatively few species that inhabit forests, adults live in pairs or alone year-round, whereas in species that occupy open habitats (snowcocks, red-legged partridges, common quail), coveys tend to amalgamate during winter months.

TRUE PHEASANTS, PEAFOWL, AND JUNGLE FOWL

All of the species in this group, apart from the tragopans, are ground-dwelling forest birds; and with the notable exception of the Congo peafowl *Afropavo congensis,* they all inhabit Asia. All males are spectacularly colored and adorned with a rich variety of vivid plumes for use during elaborate courtship displays. The crested argus *Rheinartia ocellata* has the longest tail feathers in the world. While each species has its own remarkable courtship routine, perhaps the most impressive is that of the great argus pheasant *Argusianus argus.* The male clears a hilltop in the forest as a dance floor. From here he gives loud cries soon after first light to attract females. When a potential mate

Morten Strange

appears he begins to dance around her, suddenly throwing up his wings into two enormous fans of golden decorated "eyes" which appear three-dimensional. The display culminates in an attempt to mate, after which the female will go off to nest and bring up a family, the male playing no further part. Most pheasants are either polygynous (one male mating with several females) or promiscuous. In some species, such as the ring-necked pheasant *Phasianus colchicus,* males defend a group of hens as a harem, forming bonds with them until the eggs are laid. This social system, which also occurs in the jungle fowl, is very rare in birds although it often occurs in mammals.

Pheasants have a close association with humans. A number of the most beautiful species have been released by (or have escaped from) collectors around the world. Feral populations of the golden pheasant *Chrysolophus pictus* and Lady Amhurst's pheasant *C. amherstiae* have established themselves in Britain.

The three peafowl species are famous for their train of decorated feathers, raised and fanned during courtship. Blue peafowl *Pavo cristatus* have adapted easily to living with humans, not only in

▶ *Reeves's pheasant is restricted to hill forests in central China. For many centuries its plumage (especially the tail feathers) was used by the Chinese as a decorative, ceremonial, or religious motif, a fact noted by Marco Polo in his writings.*

their native India—male peacocks and female peahens are also found in parks throughout Europe. The green peafowl *P. muticus,* however, is threatened by hunting and the destruction of its forest habitat in Southeast Asia. The Congo peafowl was discovered only in 1936 by J.P. Chapin, in the equatorial rainforest of what is now eastern Zaire, Central Africa. How this one species became isolated from all other members of the pheasant group which live in Asia is a real mystery.

The closest association between any bird and humans occurs with the jungle fowl. Chickens are

▼ *Tragopans are a group of five species restricted to high altitudes in the Himalayas from Kashmir to central China. They live in bleak, damp forests and spend much of their time in trees rather than on the ground. Males are characterized by rich red plumage, but females are dull brown. Pictured is the satyr tragopan of Nepal.*

Belinda Wright

▶ *The vulturine guineafowl inhabits arid scrublands in East Africa, but six other species of guineafowl occur across Africa in most habitats from dense rainforest to open savanna. In guineafowl the sexes do not differ obviously in appearance.*

kept wherever people live, including on the high seas. The red jungle fowl *Gallus gallus*—which still lives in the wild in India and Southeast Asia—was domesticated at least 5,000 years ago and since then has provided food (eggs and meat) to almost every human race on earth.

GUINEA FOWL

The seven species of guinea fowl all inhabit Africa, mainly in semi-arid habitats, sometimes in forests. They have largely naked, pigmented heads, with wattles and usually a crest or casque on top. Variation in the size and shape of these adornments, and the extent of naked skin, is believed to help the birds to regulate their brain temperature in different climates. All species are monogamous but congregate in flocks outside the breeding season. They feed opportunistically on the ground but roost, whenever possible, in trees. The most widespread species is the helmeted guinea fowl *Numida meleagris*. It is the only species that naturally occurs outside the Afrotropics—in both Morocco and the Arabian peninsula. Outside the breeding season, flocks of up to 2,000 birds gather towards dusk at waterholes; during the day, smaller groups regularly walk 30 to 50 kilometers (20–30 miles) while foraging.

MICHAEL R. W. RANDS

HOATZIN

The hoatzin of tropical South America is an unmistakably prehistoric-looking bird. Superficially, an adult looks like an elongated pheasant: long-necked, long-tailed, and with a long, frizzled rufous crest. The body is predominantly brown above and rusty-red to reddish-yellow below. The face is bright blue with a prominent red eye. It has an extraordinarily large crop which is a highly specialized adaptation for grinding its food —the leaves, buds, fruits, and flowers of only a few species of tree and shrub. Breeding occurs year-round. A few days after hatching, the chicks leave the nest and are able to clamber in the trees using their feet, bill, and specially adapted wings that each have two "claws" on them. These claws disappear as the chick grows, and they are absent from adults.

This remarkable species is in its own order. Its bizarre appearance has led to speculation that it provides the "missing link" between the reptiles and the birds, especially since the claws of the chick are quite similar in form to those of *Achaeopteryx*, one of the earliest known bird fossils. However, the highly adapted crop and other specializations suggest that it is not a primitive species at all, and even the claws may in fact be the product of recent evolution and not a relic.

D. & R. Sullivan/Bruce Coleman Ltd

The hoatzin is restricted to flooded forests along the banks of quiet rivers in Amazonia. For much of the year it lives in groups of 10–20 or more, but during the rainy season smaller groups split off to breed.

CRANES & THEIR ALLIES

The gruiforms are predominantly ground-living birds that have a much greater propensity for walking and swimming than for flight. Given the isolation of an island habitat, some have become flightless. They usually nest on the ground or on platform-nests in shallow water, and the young are active immediately after hatching. Most have loud vocal displays, and in many species the male and female duet. Many have elaborate nuptial dances. From a ground-loving shorebird ancestor, the 12 families of gruiforms have radiated into a diversity of niches around the world.

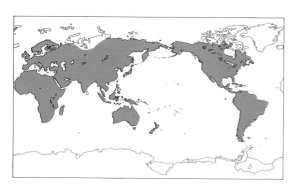

aberrant in that it has a flute-like voice and, at least in captivity, does not breed until it is five to seven years old. In contrast, other cranes have raucous trumpet-like calls and can breed in their third year.

Cranes are symbols of long life and good luck in

FLIGHTLESS ANCESTORS

Among the early gruiforms the gigantic *Diatryma* of the northern continents and *Phororhacos* of South America stood more than 185 centimeters (6 feet) tall, were flightless but possessed powerful legs and hooked bills. *Diatryma*'s skull was almost as large as that of a horse, and *Phororhacos* was believed capable of chasing down fast-running mammals. Their closest surviving relatives are the two species of long-legged and flight-capable grassland birds of South America, the seriemas. Other ancient flightless gruiforms perhaps gave rise to ostriches; shorebirds of the order Charadriiformes are believed to be the closest order to the Gruiformes. The oldest species of living bird is the North American sandhill crane *Grus canadensis* whose bones are known from deposits dating back some 10 million years. Other species of now-extinct cranes, rails, and bustards date back 40 to 70 million years.

CRANES

Found in North America, Europe, Asia, Africa, and Australia, the crane family (Gruidae) includes 15 species of large wading birds. Cranes are best-known for their loud calls, spectacular courtship dances, monogamy, and the care they lavish on their young. The crowned crane *Balearica pavonina* of Africa is considered to be closely related to the ancestral stock that subsequently gave rise to the 14 other species. Within this latter group, the Siberian crane *Grus leucogeranus* is somewhat

▶ The distinctive crest and the habit of roosting in trees are but two of the features that set the crowned crane apart from other cranes.

Stanley Breeden

▲ Close to extinction throughout much of Southeast Asia, the sarus crane is the subject of a vigorous program of reintroduction.

▶ Veiled by falling snow, a group of red-crowned cranes bugle in concert. Proclaimed as a "special national monument", a population wintering in Hokkaido, Japan, is gradually rebuilding its numbers from a low of 33 in 1952.

period. Cranes of the northern continents migrate thousands of kilometers between breeding and wintering areas. Juveniles learn the route by accompanying their parents throughout the autumn migration.

The destruction of their wetland and grassland habitats—as well as hunting on the northern continents and, more recently, poisoning in Africa—have eliminated these magnificent and space-demanding birds from much of their former range. Seven species are listed as endangered. The rarest is North America's whooping crane *Grus americana*, which in 1941 was reduced to just 15 individuals in the wild. Canada (where the cranes breed) and the USA (where they winter) have cooperated on conservation measures and there are now 181 whooping cranes in the original migratory population. There are also 64 individuals in a nonmigratory experimental population in Florida, and about 100 in captivity. Similarly, Asian nations are working together to help five species of endangered migratory cranes. Captive propagation centers for cranes have been established in China, Germany, Japan, Thailand, the USA, and Russia. Programs are underway to use captive stock to reintroduce sarus cranes *Grus antigone* into Thailand and to bolster the western population of Siberian cranes. Cranes that have been reared in captivity in visual and vocal isolation from human keepers, manipulated through hand-puppets and crane-costumed people, readily join wild cranes and thrive.

If protected, most species of cranes are remarkably adaptable to surviving in close proximity to humans. The tallest crane, the sarus of the Indian subcontinent, is protected by Hindus and thrives in densely populated areas, breeding at the edge of village ponds and foraging in the agricultural fields. Mongolians protect the demoiselle crane *Anthropoides virgo,* and on the vast grasslands the cranes often nest and forage near settlements. Likewise in regions of Africa where the crowned crane is protected, cranes roost at night on trees within the villages.

The most abundant species is the North American sandhill crane. More than half a million birds gather each March along 65 kilometers (40 miles) of the Platte River in central USA before continuing north to their northern breeding grounds. Thousands of tourists travel to Nebraska in spring to see one of Earth's greatest wildlife shows: the staging of the hordes of sandhill cranes.

LIMPKINS AND TRUMPETERS
Closely related to cranes are the limpkins of the family Aramidae, and the trumpeters of the family Psophiidae. Both families are native to the New World. There is but a single species of limpkin, *Aramus guarauna,* and it is found in the tropics. It uses its long curved bill to extract snails from their shells.

the Orient, and the attractive red-crowned crane *Grus japonensis* is prominent in their art. Cranes are, in fact, long-lived birds—one captive Siberian crane died at the age of 83 years, having successfully fathered chicks in his 78th year!

A crane pair will defend a large acreage of wetland and grassland as their breeding territory. Intruders are buffeted by trumpeting duets of the defending pair. In a secluded area, a platform nest is constructed in shallow water, and typically two eggs are laid. Both sexes assist in incubation and care of young; then the crane chicks remain with their parents until the onset of the next breeding

Jeff Simon/Bruce Coleman Ltd

▲ *The limpkin feeds almost entirely on large water snails, especially apple-snails of the genus* Pomacea. *It favors forested swamps but also sometimes occurs in marshes and scrub. In some parts of its range it is popularly known as the "crying bird" because of its eerie nocturnal wails and screams.*

In contrast to the water-loving limpkin, the three species of trumpeters are forest birds with soft feathers and short, somewhat curved bills. Like the crowned crane, trumpeters roost in trees and have velvet-like feathers on their heads and crane-like courtship dances. However, trumpeters are found only in South America, whereas the crowned crane is endemic to Africa.

RAILS

Perhaps the best known of the gruiforms is the family Rallidae, including 133 species of rails, gallinules or moorhens, and coots, although 14 species are probably extinct. Of all living groups of birds, the rails are the most likely to lose the ability to fly. One-quarter of the rail species endemic to islands cannot fly, and the introduction of rats, snakes, and domestic cats have brought about their demise. Captive breeding has been helpful for several species, and as predation and habitat problems are resolved, reintroductions have been possible.

One of the largest and most unusual members of the family is the takahe *Notornis mantelli,* a flightless purple bird from New Zealand's Fiordland. It was known from fossils, so the discovery of living birds was a great surprise. But grazing competition from introduced deer, and predation by introduced stoats, have reduced the takahe to perhaps fewer than 180 birds. Captive-

bred birds have been released and are now breeding on Maud Island.

Although many members of the Rallidae are threatened, others are remarkably successful. Almost every major wetland on all continents and many islands has a species of rail, gallinule, or coot. The New Zealand weka *Gallirallus australis,* when introduced to other, smaller islands, inflicted serious damage to vegetation and through predation brought about the demise of a petrel.

BUSTARDS

The family Otididae includes 22 species of long-legged, short-toed, broad-winged birds of the deserts, grasslands, and brushy plains in the Old World. The majority of the bustards are native to Africa; five species occur in Eurasia, and one is found in Australia.

Predominantly brown plumage provides cryptic coloration, and when alarmed, bustards often crouch and are difficult to see. Males are usually much larger than females and have elaborate ornamental plumes used in display. Their nuptial displays are bizarre—the males in some species inflate gular sacs and elevate their elongated and predominantly white neck feathers. To attract females, the male florican bustard *Houbaropsis bengalensis* of India and Southeast Asia makes helicopter-like vertical flights within its breeding territory. One male will mate with many females,

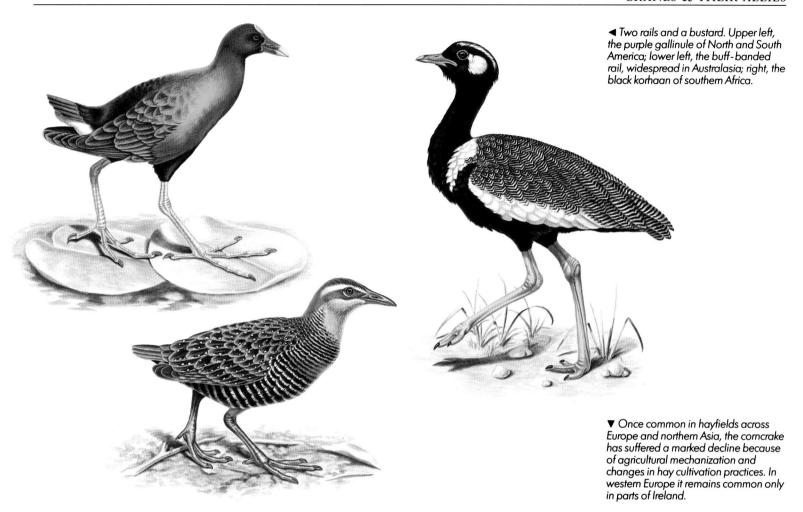

◄ Two rails and a bustard. Upper left, the purple gallinule of North and South America; lower left, the buff-banded rail, widespread in Australasia; right, the black korhaan of southern Africa.

▼ Once common in hayfields across Europe and northern Asia, the corncrake has suffered a marked decline because of agricultural mechanization and changes in hay cultivation practices. In western Europe it remains common only in parts of Ireland.

and incubation and rearing duties are usually the sole responsibility of the hen. Three to five speckled dark eggs are usually laid in a scrape on the ground.

Although loss of grassland nesting habitat to agriculture is the primary problem for the conservation of bustards, falconry is also a threat to several species. Falconry is a cherished tradition in several Arab cultures, and as a probable consequence bustards have been wiped out in much of the Middle East. Parties of falconers have recently traveled to India, Pakistan, and eastern Asia to continue their age-old sport. Concerned to keep bustards available for hunting, an elaborate captive-breeding center for bustards has been set up in Saudi Arabia. Unfortunately, bustards are difficult to breed in captivity.

In Hungary, before a bustard nest is plowed under, farmers collect the eggs for a government-supported hatching and rearing center. Later in the summer, juvenile bustards are released from rearing pens, and they join the local flocks of wild birds before migration.

UNUSUAL GRUIDS
Finfoots

The three species of finfoots (family Heliornithidae) are aquatic birds of the subtropics and tropics: one species in Africa, one in Southeast Asia, and one in Central and South America.

Richard T. Mills

Rod Williams/Bruce Coleman Ltd

▲ *The crested seriema and one other species constitute the family Cariamidae, restricted to South America. Seriemas are terrestrial, inhabit savanna, arid scrub, and thorny woodland, and feed on a variety of small animals, including snakes.*

▼ *Sunbitterns forage inconspicuously along the margins of forest rivers and streams in Central and South America. The sexes are alike in plumage.*

Generally grebe-like in external appearance, the finfoots have an unusual mixture of features that suggest affinities to cormorants, darters, ducks, coots, and rails. Drably colored and equally agile on land and in the water, these little-known birds frequent brushy edges of lakes and streams. They perch on branches over the water, seldom fly, but flap across the water in coot-like manner when disturbed. Two to seven cream-colored eggs are laid in nests built in the reeds or on low branches over the water.

Sunbittern

In the family Eurypgidae, the single species of sunbittern *Eurypyga helias,* like the finfoots, prefers heavily vegetated zones around streams and ponds. Found only in Central and South America, it is a teal-sized bird with intricately barred, soft brown, gray and black plumage, a graceful neck, and light and graceful flight. Nuptial behavior includes an elaborate display: while perched at a conspicuous spot on a branch, it spreads its wings and tail rigidly to form a continuous fan-like bank of attractively-banded feathers. Sunbitterns construct a bulky nest of sticks and mud in a tree, and both sexes assist in the incubation of the two or three nearly oval, blotched brown or purplish eggs. The young, although active immediately after hatching, remain in the nest for several weeks.

Kagu

The kagu *Rhynochetos jubatus* (sole member of the family Rhynochetidae) is found on the island of New Caledonia in the Pacific Ocean and is considered to be closely related to the sunbittern. It is slightly larger than the domestic chicken but with longer legs. It is incapable of flight but is able to glide. Pale gray with an elaborate crest, the kagu is a forest-floor bird. Like cranes, the male and female kagu have a loud and elaborate duet, and they engage in bizarre courtship displays in which they whirl around holding the tip of the tail or of a wing in the bill. They build a stick nest on the ground and produce a single egg. Both male and female assist in rearing the young. Eggs and young have fallen prey to introduced cats, dogs, pigs, and rats, and much of the kagu's forest habitat has been destroyed by humans in the course of farming and open-pit nickel-mining. It now survives only in a few inaccessible mountain valleys.

THE TINY GRUIDS
Mesites

The three species of mesites (family Mesitornithidae) of Madagascar are an ancient group of isolated lineage. Superficially resembling large thrushes but with the gait of a pigeon and the behavior of a rail, mesites are forest-floor species that seldom fly. The stick nest is constructed on a

Michael Fogden/Bruce Coleman Ltd

Australian Picture Library

low branch, two or three eggs are laid, and there is some evidence of polyandry (the female mating with two or more males in the breeding season). As the forests in Madagascar continue to be destroyed, and if their primary egg-predator, the brown rat, continues to flourish, the mesites can be expected to decline in numbers.

Hemipode-quails
Polyandry and a three-toed foot are salient features of the hemipode-quails or buttonquails (family Turnicidae). With 15 species widely distributed in southern Asia, Africa, and Australia, these terrestrial birds, although true gruiforms, have some similarities with sandgrouse and pigeons. Females are larger and more brightly colored than males, and they have a loud booming call. They circulate among several males, and they are particularly aggressive in expelling other females from their breeding territory. Both sexes build the

ground-nest, but only the male incubates the clutch of four eggs. The incubation period is about 12 or 13 days, and within two weeks of hatching the young are capable of flight. First breeding can take place when hemipode-quails are only three to five months old.

Collared hemipode
The collared hemipode or plains wanderer *Pedionomus torquatus* (sole member of the family Pedionomidae) inhabits Australia's open grasslands and plains. It resembles its relatives the hemipode-quails, but its persistent hind toe, egg shape, and paired carotid arteries suggest it deserves family status. The bird is about 16 centimeters (6¼ inches) long, and the female's chestnut-colored breast distinguishes her from the male. The male incubates the four-egg clutch and rears the young. They seldom fly and prefer to run and hide when afraid.

GEORGE W. ARCHIBALD

▲ *The little buttonquail inhabits grasslands in inland Australia, becoming rare at higher altitudes and toward the coast. As in other buttonquail species, the male assumes sole responsibility for incubation and care of the young.*

ORDER CHARADRIIFORMES
• 14 families • 82 genera
• 292 species

SMALLEST & LARGEST

Least sandpiper *Calidris minutilla*
Total length: 11 cm (4½ in)
Wingspan: 33 cm (13 in)
Weight: 23–37 g (¾–1⅓ oz)

Great black-backed gull *Larus marinus*
Total length: 64–78 cm (25–30 in)
Wingspan: 150–163 cm (5–5⅓ ft)
Weight: up to 18.7 kg (41 lb)

CONSERVATION WATCH
!!! The 4 species listed as critically endangered are: black stilt *Himantopus novaezelandiae*; Eskimo curlew *Numenius borealis*; slender-billed curlew *Numenius tenuirostris*; Chinese crested tern *Sterna bernsteini*.
!! There are 8 species listed as endangered: Chatham Islands oystercatcher *Haematopus chathamensis*; Jerdon's courser *Rhinoptilus bitorquatus*; New Zealand dotterel *Charadrius obscurus*; St Helena plover *Charadrius sanctaehelenae*; shore plover *Thinornis novaeseelandiae*; Tuamotu sandpiper *Prosobonia cancellata*; Nordmann's greenshank *Tringa guttifer*; Saunders's gull *Larus saundersi*.
! 22 species are listed as vulnerable.

WADERS & SHOREBIRDS

The shallow waters of the world—pools, puddles, and the shores of lakes, rivers, and sea—seem always to have been rich in tiny creatures. They are therefore valuable habitats, and a whole order of birds, the waders or shorebirds, Charadriiformes, has evolved to exploit them. These places do, however, have some fundamental limitations: birds must wade or swim after food; and the shallow waters are vulnerable to rapid change, drying-up in hot areas, freezing in cold ones, so the birds must be able to fly far and fast should the need arise.

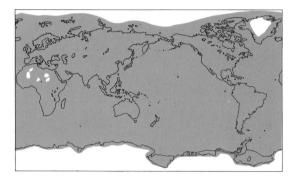

BIRDS OF COASTAL AND INLAND SHORES
This order of birds comprises several groups and subgroups that have diverged to take advantage of different resources. The typical waders are perhaps those of shallow waters and shores. Their legs are long and slender for wading, necessitating long slender necks and bills for feeding, and their wings are long and narrow for fast flight. Species that prefer estuaries and seashores tend to be a little heavier in build, with a wider range of bill sizes and shapes for snatching, probing, or prizing the more varied prey.

Gulls, a smaller and more uniform group, are mainly seabirds, although a few species use fresh waters. Legs tend to be stouter and feet webbed for swimming, enabling them to exploit the surface of deeper water as well. Their bill is generally thicker, and hooked a little at the tip, for they are scavengers as well as predators. They spend much more time on the wing in search of food.

The terns, a small group probably derived from gulls, have shorter legs and smaller feet, but long narrow wings—flight being more important than walking or swimming. Their tail is long and usually

► *Not quite as strongly migratory as many waders, the Eurasian curlew breeds on moorland and winters on coastal mudflats and river estuaries.*

Richard T. Mills

forked for quick maneuverability in flight, and the bill is tapered, for they are mostly plunge-divers, usually at sea, exploiting deeper water layers than gulls.

Finally the auks, the smallest main group, have moved out to exploit the deeper seas, swimming underwater after their food, and in consequence becoming penguin-like. The body is compact, with webbed feet set well back, and the wings are used as underwater flippers as well as in flight. Heads tend to be largish, and bills stout and strong. While much smaller and less specialized in structure than penguins, which live in the Southern Hemisphere, auks appear to be evolving to fill a similar place in the avifauna of the Northern Hemisphere.

WADERS, OR SHOREBIRDS

Waders are generally birds of open places, many sociable in their nesting. Nests are often sketchy or non-existent, and the eggs are relatively large, producing downy young that can be active from the moment they hatch. The typical waders, of which there are some 216 species, form the major part of this order.

Curlews, sandpipers, and snipe

The family Scolopacidae includes 88 species. They tend to have cryptically colored, brown or grayish, streaked plumage, but may assume brighter orange or black patterns when breeding. They share a similar basic structure—a long body, narrow wings, and long legs—but in size they vary from the eastern curlew *Numenius madagascariensis* at about 66 centimeters (26 inches) to the least sandpiper *Calidris minutilla* at 11 or 12 centimeters (4½ inches).

There are oddities, such as the short spatulate bill of the spoonbilled sandpiper *Eurynorhynchus pygmaeus,* but most bills are narrow, varying in length and curvature. They range from the long decurved bills of curlews, which probe deep into worm burrows in mud, and long straight ones for godwits searching mud under water, to shorter probes, and the small stubby bills of stints. The bills have sensory nerve endings near the tip, enabling the birds to sense food as they probe blindly. Most bills seem designed to exploit part of the range of organisms in the mud and sand of the birds' winter quarters. On rich mudflats large

▲ The Charadriiformes constitute a diverse group. Top, the black skimmer, like other skimmers, has a highly asymmetrical bill for its habit of flying just above the water and dragging the lower bill in the water to catch food. The banded lapwing (center, below) is a common Australian plover, while the lesser golden plover (far left) migrates regularly from western Alaska and Siberia to as far south as Australia. During the breeding season the bill of the male Atlantic puffin (far right) develops bright colors.

▶ It is characteristic of the Scolopacidae family that various species associate freely together on their wintering grounds, especially at high-tide roosts. At least three species are portrayed at this roost in northwestern Australia: the bird stretching its wings is a black-tailed godwit; most others of equal size are bar-tailed godwits, while the smaller birds are great knots.

Brian Chudleigh

flocks can feed together, and may move and maneuver in close rapid flight.

Almost all of these species nest in the Northern Hemisphere, many using the short summer of the Arctic tundra when waterlogged ground overlays the permafrost. The breeding grounds may not seem ideal for the adults' bill-shapes, but they have abundant insects that soon become food for the short-billed young. The open areas used by many for nesting encourage aerial displays, which are accompanied by the musical piping calls typical of these birds, and the dark-bellied breeding plumage of some species may be a visual advantage. The eggs, usually three or four, are often laid in bare scrapes—no nest material is carried, and only what is nearby is used. The downy young usually feed themselves from the earliest stage, with a few exceptions. Snipe and woodcock have proportionally the longest bills, and probe deep into the soft mud of inland marshes. The large laterally-placed eyes with almost panoramic vision watch for predators, not prey. These species bring food to their short-billed chicks.

The breeding season may be short, and when it ends the birds migrate south to other shallow waters or coasts, some even going as far as Australasia. Although often territorial when breeding they may be sociable at other times, with loud calls for maintaining contact.

Within this family there are many subgroups—and vernacular names such as whimbrel, redshank, greenshank, yellowlegs, willet, tattler, woodcock, dowitcher, surfbird, knot, sanderling, ruff—and many have specialized adaptations. The two turnstones (*Arenaria* species), with small plump bodies, stubby bills and strong necks, find food by flipping over stones and seaweed. The seemingly delicate phalaropes (*Phalaropus* species) swim and rotate to pick tiny creatures from the water's surface, and two species spend the winter far out at sea; the third, Wilson's phalarope *P. tricolor,* winters around South American inland waters.

In the phalaropes, and in some other families of waders such as jacanas, painted snipe, the Eurasian dotterel, and some sandpipers, the female is larger than the male, more brightly colored and dominant. She is polyandrous, leaving several males to care for clutches of eggs and the subsequent young. The ruff *Philomachus pugnax* goes to the other extreme: the breeding male grows a great individually-colored and patterned ruff, and ear-tufts, and these are displayed in ritual posturing on a communal "lek" or display ground where females mate with the most conspicuous and impressive males.

Oystercatchers

The oystercatchers, family Haematopodidae, are birds of shores, estuaries, and stony rivers. They are heavier in build, and the long bill has a blunt laterally-flattened tip that opens shellfish and chips limpets off rocks, as well as taking other prey. They are pied or black, loud-voiced, and aggressive. They bring food to their small young.

Plovers and dotterels

The 66 species of the family Charadriidae also show bold and bright color patterns. They have short, stout bills and feed in a "run and snatch" fashion on visible prey. They need rather bare open habitats, so coastal beaches and bare watersides support many smaller species. The golden and gray plovers winter on coasts, but nest on tundra or bare northern uplands. The boldly-colored and broader-winged lapwings and wattled plovers are birds of moister grasslands. The habitats of the others vary from the semi-deserts of Australia's interior to the Arctic/Alpine mountaintops of Eurasia.

They are territorial when breeding, usually with conspicuous aerial displays—the Eurasian lapwing *Vanellus vanellus* being famous for its erratic broad-winged tumbling.

Richard T. Mills

One small plover, the New Zealand wrybill *Anarhynchus frontalis,* is unique among birds in having a bill that bends sideways, to the right. It appears to be an adaptation to extracting insects from undersides of stones in riverbeds.

Thick-knees, or stone curlews

The stone curlews, family Burhinidae, are like very large, round-headed, and large-eyed plovers of

▲ Oystercatchers of one species or another are found along seashores almost world-wide. The plumage of some species is entirely black; others are boldly pied, like this Eurasian oystercatcher.

Klaus Wernicke/Silvestris GmbH/Frank Lane Picture Agency

◄ Outside the breeding season the male ruff looks little different from the female (known as a reeve) except for his substantially larger size. In spring, however, males don a cape or "ruff" of long loose feathers around the neck and on the crown. This finery is used in courtship displays and differs widely in color and pattern among individuals. The color varies from pure white through a range of reds and browns to black; the feathers may be barred, spotted or plain, and no two birds are precisely alike.

Belinda Wright

bare inland places and stony riverbeds. Cryptically striped, they tend to rest by day and are active at dusk and by night when their big staring eyes, pied wing-patterns, and loud wailing and whistling calls become apparent. The seven small *Burhinus* species have short stout bills, but in the two larger *Esacus* species these become longer and more massive. The beach thick-knee *E. magnirostris,* of Oriental and Australasian coasts, has the largest bill and feeds on crabs and marine invertebrates, living and nesting on the shore.

Crab-plover

The crab-plover *Dromas ardeola* is like a very large plover or stone curlew, with a massive bill as long as the head. It has a conspicuous black and white plumage and is a shorebird of Indian Ocean coasts and islands, feeding on crabs. It breeds in colonies on sandy areas, digging a tunnel more than 1.5 meters (5 feet) long, and laying one white egg. The young remains in the tunnel while the adults bring food, and it is still fed after it leaves the nest. Adults fly slowly and low, often in chevrons, with the head supported on the shoulders.

The adaptive radiation of the waders that produced these various groups of birds has also resulted in the evolution of more-specialized small groups variously adapted to wetter conditions, to deserts, to mountains, to aerial hunting, or to life on the Antarctic borders.

Ibisbill

The ibisbill *Ibidorhynchus struthersi,* a stoutish gray bird with black on the face and breastband, is of

▲ *A Eurasian stone-curlew. Members of the Burhinidae are widely known as stone-curlews in Europe, thick-knees in North America, and dikkops in Africa. They have large yellow eyes, an oddly furtive manner, and loud wailing cries uttered usually at night. Several species inhabit arid scrub-lands.*

▶ *The crab-plover is restricted to tropical shores of the Indian Ocean, where it frequents sandy beaches and mudflats, feeding largely on crabs. Compared to other waders, its breeding habits are extraordinary: it breeds in dense colonies in sand dunes, digging burrows in which the single, pure white egg is laid. The young are dependent on their parents for food and, as in the oystercatchers, this dependence persists to some extent even after fledging.*

P. Evans/Bruce Coleman Ltd

Geoff Longford

doubtful affinity, tentatively linked with both stilts and oystercatchers. Its heavy build, shortish legs, and curved bill may be adaptations for feeding in the stony mountain streams where it probes for food. It is confined to southern Asian uplands.

Painted snipes

On lower wetlands there are two painted snipe (genus *Rostratula*), already mentioned for their sex-role reversal. They are birds of tropical lowland swamps, one species in the Old World, the other in South America. They have slightly curved long bills and broad wings, and feed on insects and worms, nesting within the swamps. They are furtive and rather silent birds, active at dusk and dawn, and as a result little is seen of the beautiful intricately-patterned plumage that is shown in flight and in spread-wing displays.

Stilts and avocets

Stilts and avocets, family Recurvirostridae, are more cosmopolitan in distribution and occur around more open waters. They are slender and long-legged, and unlike other waders they have webbed feet. The banded stilt *Cladorhynchus*

leucocephalus has the smallest range, on saline waters of southern Australia. Black, chestnut, and white, with long thin legs and bill, it feeds in shallow waters, mainly on brine shrimps; it is nomadic to take advantage of rain at inland lakes, and at suitable sites will form large breeding colonies of shallow scrapes near water. The black-winged or pied stilt *Himantopus himantopus* is seen on every warm continent, in forms varying in the black coloration on the head and neck. The neck is long; the bill long, thin, and straight; and the legs so long that it must tilt its body to reach the ground. It looks absurd, but when wading belly-deep in water, picking food from the surface, appears natural. It is a freshwater species, nesting in loose colonies, and making a quite substantial nest on a tussock in the water. It has a yelping call and flies with long legs trailing behind. The feet are only slightly webbed.

The avocets (genus *Recurvirostra*), one species on each warm continent, have pied plumage with heads varying from black and white to chestnut, buff or all-white. Their very thin, long bill is upturned, and they feed in shallow brackish waters, sweeping the bill from side to side in search of tiny prey.

▲ As a breeding bird, the banded stilt is characteristic of the most inhospitable deserts of Australia, and few nesting colonies have ever been found. In its behavior it shows some remarkable similarities to the flamingos: like them, it is strongly gregarious, favors arid salt lakes, feeds largely on shrimps, often swims as it feeds, and breeds erratically in very large dense colonies.

► Like other jacanas, the lotusbird or comb-crested jacana prefers deep, permanent lagoons with an abundance of floating vegetation. This species occurs in Indonesia, New Guinea, and Australia; others inhabit tropical wetlands around the world. During the breeding season the birds are strongly territorial, but they often congregate in flocks at other times of year.

Belinda Wright

Jacanas

The family Jacanidae has gone a stage further as waterbirds. There are five species in warmer climates, rail-like with rounded wings and sharp bills; four have forehead plates or wattles. The large legs are long and strong, with long splayed long-nailed toes that can support the bird's weight as it runs over floating vegetation or waterlily leaves. They live on still inland waters. The skimpy nest rests on water plants. The adults can move the boldly-scribbled eggs or downy young a little way by gripping them under the closed wings.

Coursers and pratincoles

The desert waders are classified as "coursers" in the family Glareolidae. The little Egyptian plover *Pluvianus aegyptius* is an odd species of river sandbars and banks. It is boldly patterned, and ploverlike in appearance and feeding. With rapid kicking, it buries its two to three eggs in warm sand for incubation, and cools them with water from wet belly feathers; it cools chicks in the same way and buries them too if danger threatens. The eight coursers are larger birds with long legs and short toes, and small curved bills. They occur in dry country and desert borders of Africa, Arabia, and India. They are well camouflaged and can run fast, tending to crouch in alarm, but able to fly well and for long distances.

This family also contains the aerial waders, the pratincole species, which occur through warmer parts of the Old World. These show some swallow-like adaptations for catching insects on the wing, although they can also run fast on long legs. They have more pointed wings, and forked tails, and the bill is short but with a very wide gape. They live on large rivers or marshes and are highly gregarious, feeding in flocks with constant high-pitched calls. Their flight is swift and agile. The Australian pratincole *Stiltia isabella* has longer wings and legs and is a migrant, nesting in inland gibber (stony) deserts within reach of water.

Seedsnipes
High-latitude grassland and lower southern plains in South America support the four species of seedsnipe (family Thinocoridae). These are curious gamebird-like waders resembling partridge or quails. They are plump, mottled brown, with short stout legs and short deep bills. Their flight is snipelike and they feed on seeds and plants.

Sheathbills
Least wader-like are the two sheathbills (*Chionis* species) of the Antarctic shores. Like large white pigeons or small domestic fowl, they have short strong legs and run well. Their flight is pigeon-like but lazy. The posture is upright, and they have a rounded head and a very stout deep bill, with bare wattles around the bill-base and the eyes. They are scavengers, haunting the breeding colonies of penguins, cormorants and seals, taking eggs, carrion and feces, and at other times feeding on small shoreline creatures or around rubbish dumps. They nest in scattered pairs in crevices and rock cavities and usually rear two young.

GULLS AND GULL-LIKE BIRDS
Skuas
The skuas or jaegers (family Stercorariidae), consist of seven gull-like species, but are brown in color with white wing-patches. The largest species (genus *Catharacta*) are like big, heavily built, broad-winged gulls. Far-ranging, they breed on shores of cold seas, with three in the south to Antarctic regions, and one in the North Atlantic. They scavenge around seabird colonies, taking eggs and young, and killing weak birds. Strong and fast on the wing, they are piratical as well as raptorial,

forcing other species to disgorge food.

The three smaller species (genus *Stercorarius*) are more slender and long-winged, with the two central tail-feathers of adults elongated, tapering or rounded at the tip. They breed on the Arctic tundra but at other times range widely over the seas into the Southern Hemisphere. They are fast-flying and very agile, and find much of their food by harrying smaller birds such as terns, forcing them to disgorge fish in aerial attacks. When breeding they also hunt lemmings and voles, kill small birds, and take carrion.

Gulls
The gulls of the family Laridae are a readily recognizable group, solidly built, with long strong webbed-footed legs for running and swimming, and long wings for the steady sustained flight that goes with aerial scavenging. Buoyant gliding and soaring is often involved, as food is hunted over water as well as land.

With a few dusky exceptions, the plumage is mainly white with gray or black across the back and wings. There is little conspicuous patterning, but some species have a brown or black hood that combines with a red or yellow bill to act as a display pattern. Color and patterns are similar in both sexes, and young birds have mottled brown plumage with dark bills and a dark tail band.

The bill is slender in the smaller inland species that feed mainly on insects, but in larger species it is stout with a tip for tearing —adapted for feeding on carrion, particularly fish—but gulls will feed on small marine life exposed by tides or take any small creatures, and the larger gulls will attack and kill young or smaller seabirds. (They are probably responsible for the nocturnal nesting habits of many smaller seabirds.) Gulls tend to be sociable in

Jean-Paul Ferrero/Auscape International

▲ *The yellow-billed sheathbill inhabits Antarctica and several subantarctic islands.*

▼ *A southern skua raids a gentoo penguin colony. Agile, strong, and piratical, skuas take the place of raptors on many subantarctic islands.*

Francisco Erize/Bruce Coleman Ltd

Geoff Longford

▲ Many calls and postures of gulls around the world are highly ritualized, and differ only in detail from species to species; this feature is useful in unraveling their relationships. These kelp gulls are uttering the so-called "long call", common to many of the larger gulls. The kelp gull occurs in Australia and New Zealand, South Africa, and southern South America.

nesting, resting, and feeding; and they respond quickly when the behavior of individuals indicates that food has been found.

The 48 species are spread around the shorelines and islands of the world, and to the larger inland waters. Food is taken from land or from the surface of the water, and smaller species sometimes hunt insects on the wing. As scavengers they have readily learnt to use discarded waste from human activity, following boats, searching recently plowed land, and feeding on rubbish tips.

Nesting is usually colonial, using sites varying from cliff ledges to level ground, and from the Chilean desert to polar shores. The small North American Bonaparte's gull *Larus philadelphia* is an exception, nesting inland in conifer trees. A gull's nest is usually a substantial cup of carried vegetation, normally with two or three spotted eggs. The dark-mottled downy young are fed by the adults, but are mobile and even when feathered may run and hide if threatened. On the nests of kittiwakes (*Rissa* species), which are trampled drums of mud and plants fixed to small projections of precipitous cliffs, the young are adapted to making little movement on the site.

The voice of gulls is usually harsh—wailing, cackling or sqawking calls—and colonies are noisy.

Terns

Terns are gull-like but adapted for more specialized feeding such as plunge-diving. The plumage tends to be pale gray and white, with black on the crown or nape, but a few species are mainly brown or black. Young are brown-mottled on the back and wings. Wings are long and slender, the tail distinctly forked, the bill strong, thin and tapering, and the legs rather small and short, with webbed feet.

Perhaps the least modified are *Chlidonias* species: the two black terns and the whiskered tern. Mainly black or gray when breeding, they have thin bills and less forked tails. These so-called "marsh terns" are birds of inland waters of warmer climates and are very light on the wing, feeding mainly by briefly dipping to snatch tiny creatures from the surface. They will also hunt insects over land. They have sharp, short, high-pitched call-notes. Their nests are in colonies on inland marshes and lakes, flimsy structures usually on floating vegetation, sometimes in shallow water. Adults bring food, but after a few days the active downy young may leave the nest to hide in nearby vegetation. After breeding they disperse rather than migrate, keeping mainly to inland waters.

The typical sea terns, 34 species, have compact and streamlined bodies, a heavier head with a tapering dagger bill, long and narrow wings, forked tails for quick braking and maneuvering, and relatively small legs and feet, since the birds rely mainly on flight. Most are white with gray backs and wings, and a black cap or nape when breeding, but a few are brown or black on the upper parts and head, with a white forehead. The Inca tern *Larosterna inca* differs in being deep gray with an elongated and exquisitely curled mustache, and is also peculiar in nesting in cliff cavities. In size, the sea terns range from the big Caspian tern *Sterna caspia,* about the size of some larger gulls at 53 centimeters (21 inches), to the little tern *S. albifrons* at 24 centimeters (9½ inches). They fly low over oceanic or inland waters, spotting prey, hovering briefly, and then plunging for food, often sending up a plume of spray on impact. The sooty tern *S. fuscata* may snatch small fish or squid at or above the surface when larger fish pursue them, and may fish at night.

Sea terns are highly sociable nesters, making a collection of scrapes, unlined unless material is to hand, on beaches, sandbanks, and low, sparsely-vegetated shores. Even where space is available the nests are closely packed, often a bill-stab apart, and there is frequent bickering. Displaying males parade around their mates with plumage sleeked and a small fish held high. Calls are usually harsh single notes, with higher-pitched alarms. Colonies are noisy and are fiercely defended by aerial attacks on intruders. Birds nesting on hot tropical beaches

may shade the nest at times rather than incubate, and return with wet belly feathers to moisten eggs or chicks. The older young of some species assemble near the water's edge to await returning adults. Migrating young leave the colonies while still relying on parents for food.

As strong fliers, sea terns spend much of their time on the wing, some species migrating long distances. The Arctic tern *S. paradisaea* is famed for traversing the world from one polar sea to the other twice yearly.

An exception among these species is the gull-billed tern *Gelochelidon nilotica*. It has a shorter, thick bill, less deeply forked tail, and longer legs than other terns. It is a bird of open inland waters, coastal lagoons, estuaries, and salt-marsh. It tends to hunt on the wing, swooping to snatch insects on land, and small fish, crabs or frogs from the surface of water, but may also hunt on foot. It nests colonially on small raised islands in shallow water.

The noddy terns consist of the three blackish-brown *Anous* species and the pale blue-gray noddy *Procelsterna cerulea,* all with whitish foreheads. The bill is slender, the wings long, the tail broader and wedge-shaped with a blunt fork at the tip, and the legs are short. Terns of tropical and subtropical seas, they are coastal and oceanic birds, mainly around islands, and generally do not range widely. They are not plunge-divers but surface-dippers, hovering to snatch from the surface or catch in the air smaller fish or squid trying to escape

▼ *A pair of Arctic terns greet each other. Arctic terns show strong fidelity both to their mates and their nest sites of previous years, but there is little evidence that they remain together during their winter migrations. Rather, Arctic terns tend to congregate at roosts immediately on their return to the breeding grounds and for a few days before the nesting colonies themselves are reoccupied; pair bonds are re-established in courtship flights at the roost and above the nesting colony.*

Townsend P. Dickinson/Comstock

C.A. Henley

lower mandible compressed laterally to a flat blade projecting beyond the upper mandible.

They usually feed by flying low over the water with the tip of the longer lower mandible plowing a little below the surface. When prey is touched, it is snapped up. The young have even-lengthed mandibles and can feed more normally. The flight is light and graceful with steady, shallow beats of raised wings.

AUKS
The 22 auk species show a certain uniformity. Medium-sized to small, they are plain-colored, varying from black or dark above and pale below to all-dark species, some with pale wing or shoulder patches. Breeding may bring vivid colors on the bill and gape, and odd feather adornments on the heads of some species; but the voices are unexceptional and calls are mostly short and deep hoarse notes.

With largish heads, thick waterproof plumage on a compact body, and a tendency to ride high in the water, they look stout and stubby when

▲ *The white tern breeds on oceanic islands in the tropical Pacific, Indian, and Atlantic oceans. It builds no nest: instead it balances its single egg in any available crevice on the upper surface of the limbs of trees.*

▶ *Weighing around 5 kilograms (11 pounds), the great auk was by far the largest of the alcids. It could not fly, and was restricted to colder regions in the northern Atlantic Ocean. It was driven to extinction by sailors and fishermen, mainly to provision boats, and the last birds were killed in Iceland in 1844.*

underwater predators. The birds will flap onto the surface in a shallow plunge, or swim to seize surface prey. They swim well, and rest on the water. They feed sociably, sometimes in large numbers, and the brown noddy *A. stolidus,* which is more migratory, may occur well out at sea in huge flocks. Noddies nest colonially in bushes or trees such as mangroves, building untidy platforms of seaweed, sticks, and other debris, but in some regions they may nest on cliff ledges or even on the ground. The white forehead is used in head-nodding displays, and they have other terrestrial and aerial posturings.

The other odd type of tern is the fairy or white tern *Gygis alba.* A tropical coastal tern, it is small, delicate-looking and pure white, slender-billed and with large black eyes. It often feeds at dusk, snatching small fish and squid at or above the surface, in similar fashion to the noddy terns. It makes no nest, balancing its single egg on a small precarious surface such as a tree branch, rock pinnacle or similarly hazardous place. The chick has strong feet with sharp claws, and from the outset must cling to stay on the site.

Skimmers
The skimmers (genus *Rhynchops*) are superficially tern-like birds, white with black wings and crown. There are three species, one each in Africa, India, and the Americas. They are highly sociable at all times, tending to feed, rest, and breed in flocks. Relying wholly on flight for feeding, and resting on the sandbars or gravel of shores and rivers at other times, they have long tapering wings, short tails with a fork, and shortish legs with webbed feet. The unusual bill is long and tapering, with the

floating or paddling at the surface. Under water they are more streamlined and move with agility, swimming with strong strokes of the wings held out from the body, and steering with the feet. The larger species take mostly fish, but the smaller ones may rely on plankton. The wings are short, small, strong, and firm-feathered for swimming. Flight is usually direct and low over the water, with continuously and rapidly beating wings. Having small tails they tend to use their splayed webbed feet in maneuvering.

The guillemots or murres—larger auks up to 46 centimeters (18 inches) long—are birds of colder and arctic seas, with narrow, tapering bills. They breed on rocky coasts. The common guillemot *Uria aalge* and the thick-billed or Brunnich's guillemot *U. lomvia* huddle close-packed on broader ledges, making no nest; the very varied color and pattern of their single egg may aid recognition by the owners. As with some other auk species, the young flutter down from the ledge before they are fully fledged and go out to sea escorted by the parents. The smaller *Cepphus* guillemots breed in deep crevices, laying two eggs, and the young leave more fully fledged.

These auks and the puffins breed in the north of both the Atlantic and Pacific oceans. Two puffins (*Fratercula* species) have stout deep bills which in breeding birds grow temporary bright bill-sheaths producing a great parrot-like bill, with wattles on the bill-base and eyelids completing a clown-like face. The tufted puffin *Lunda cirrhata* has a simpler pattern ornamented with big swept-back tufts of blond plumes. Puffins burrow into the softer soil-covering of cliffs and islands, and they gather nest material. The serrated bill-edge allows them to pack in slender fish, held crossways, to carry to the single young.

The North Atlantic razorbill *Alca torda* is a guillemot with a deep, blunt, and laterally flattened bill, resembling the common guillemot in habits. Its close relative was the great auk *A. impennis,* which used to nest in colonies on low islands in the North Atlantic; being flightless it was easily killed, and it was finally exterminated in 1844.

The smallest of the Atlantic auks is the little auk *Alle alle,* a tiny bird with big head and stubby bill. It winters out at sea except when driven inshore by storms, and nests in huge swarms in the cliff screes of arctic islands. It is more active and agile on the wing than larger species.

In the North Pacific the smallest members of the auk family are called auklets and murrelets, six species each, breeding mainly in colonies on coastal islands and archipelagos. For safety they tend to nest in burrows or rock crevices; they usually have a single egg. The auklets are the more heavily built and go farther out to sea when not nesting. Like puffins, some develop brightly colored bill-sheaths which are shed after each breeding season, and some have fine projecting feather tufts on the forehead and cheeks.

Murrelets are smaller and tend to stay closer inshore. They have small fine bills, stubby or slender. The two *Brachyramphus* species are peculiar in having mottled brown breeding

Dean Lee/Weldon Trannies

plumage, and they may fly inland to nest. Both are little known, but Kittlitz's murrelet *B. brevirostris* has been found using a nest-hollow in stony ground above the tree level on mountains. The marbled murrelet *B. marmoratus* sometimes occurs in huge numbers but only a few nests have been found: two on bare tundra; and two on large branches or lodged debris high in forest conifers, one more than 8 kilometers (5 miles) inland. The young bird must try to reach the sea in its first flight.

COLIN J. O. HARRISON

▲ *The horned puffin is among the most numerous birds along the coast of Alaska. It is named after the small fleshy appendage over each eye, which the bird can raise or lower at will.*

PIGEONS & SANDGROUSE

There are two quite different families of birds within this order, and they may not even be closely related. Pigeons and doves of the family Columbidae are basically seed- and fruit-eating, tree-dwelling, terrestrial birds, occurring throughout the world except in the high Arctic and classified in more than 300 species. Less well known are the 16 species of sandgrouse of the family Pteroclididae, which are desert-dwellers of Africa and Eurasia. Scientists argue frequently about whether they are related to pigeons at all, some suggesting they are waders (Charadriiformes), others that they should form their own order. A third family, Raphidae, now extinct, consisted of the dodo and two related species of the Mascarene Islands in the Indian Ocean.

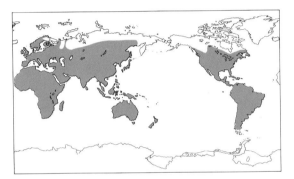

SANDGROUSE

Sandgrouse are medium-sized birds, specialized for a life in the deserts and semi-arid regions of Africa and Eurasia. They have dull, well-camouflaged brown, gray or khaki-colored plumage and are compact and streamlined, with small heads and short necks. The tops of their feet are densely clothed in feathers. Males and females have different plumages. Their adult feathers are similar to those of pigeons; and like pigeons (and

► The Nicobar pigeon (left) feeds on the ground in rainforest and mangrove swamps; it inhabits Indonesia and New Guinea. The superb fruit dove (top right) occurs in New Guinea and northeastern Australia; the pin-tailed sandgrouse (bottom right) inhabits southern Europe and North Africa.

unlike most other birds) they can drink by sucking, although less expertly than pigeons—they have to raise their heads to swallow. They differ from pigeons, however, in several important ways: they cannot produce crop milk; they have a pair of large functional ceca (pouches, or blind tubes, forming the beginning of the large intestine); they have a different syrinx or voice-box; and they have oil-glands that produce an oil which is used to preen the feathers.

They are very strong and fast fliers, the Namaqua sandgrouse *Pterocles namaqua* of the deserts of southern Africa being able to outpace a falcon in level flight. They need to drink regularly and may have to commute over 60 kilometers (37 miles) to water. A typical species is the pin-tailed sandgrouse *P. alchata* of North Africa, Spain, the Middle East, and Central Asia west of the Caspian Sea. Flocks of at least 50,000 have been recorded at waterholes in Turkey. In the breeding season, March to May, they break up into pairs, then gradually congregate again in flocks after the young fledge in September.

Sandgrouse nest on the ground, and the young leave the nest a few hours after hatching. The chicks take every opportunity to shield themselves from the hot sun and even shelter under their parents while moving. As a protection against predators they half-bury themselves in the sand under the shade of bushes. Sandgrouse have a unique way of watering their young: the male flies

to a waterhole and wades in with his feathers lifted; the central feathers soak up water, then the male returns to his family where the chicks drink from his wet feathers.

PIGEONS AND DOVES

About three-quarters of the 304 species of pigeons and doves live in tropical and subtropical regions. The term "pigeon" is used for larger species, and "dove" for the smaller, more delicately built ones. They are medium-sized to small birds which feed on seeds and fruits. Most are tree-dwelling, yet some of the common species feed in huge flocks on the ground. The plumage is soft and dense, and the feathers have characteristically thick shafts and fluffy bases. They have no, or very small, oil-glands—instead, special plumes disintegrate to produce powder that cleanses and lubricates the plumage. The most specialized feature of the family is the ability to produce crop milk: when the birds are breeding, special glands in the crops of both male and female enlarge and secrete a thick milky substance which is fed to the young.

The sexes are usually similar in appearance, but some species have different male and female plumages. Pigeons and doves have characteristic sexual and advertising displays, such as bowing, and special display flights. Their calls are usually pleasant cooing notes. All species build flimsy nests of a few sticks in trees, or on the ground or on

▲ *The African green pigeon is common and widespread over much of the African continent; closely related species also occur in tropical Asia. These pigeons are strongly arboreal and feed mainly on fruit.*

Frithfoto/Australasian Nature Transparencies

of fruit dove (genus *Ptilinopus*) of the Indo-Pacific region. These smallish plump pigeons are spectacularly colored with bright greens, brilliant reds and oranges, purples and pinks, blues and golds. They live high in the canopy of the rainforests, and some species are found only on single islands in the Pacific Ocean. Fiji, for example, has three very specialized species collectively called golden doves: the orange dove *P. victor* is fiery orange; the golden dove *P. luteovirens* is metallic greenish-gold; and the yellow-headed dove *P. layardi* is green with shining gold fringes to the feathers. Fruit doves eat only the fruits of rainforest trees and have specialized digestive systems—features they share with the 36 species of imperial pigeons (larger birds, in the genus *Ducula*) which have the same distribution. Most birds take grit that lodges in the crop and helps grind up food, but fruit doves and imperial pigeons do not take grit and have a thin gizzard with horny knobs which gently strips the flesh from the seeds. The seeds are defecated whole and so these birds act as important dispersers for rainforest trees.

The New Guinea rainforests are the home of some spectacular birds such as the three species of crowned pigeon in the genus *Goura*. The size of small turkeys, they are the largest pigeons and are characterized by big filmy crests and subtle purple and gray plumage. They forage in small groups on the forest floor and, despite their size, nest up to 15 meters (50 feet) in trees. Elsewhere in the Pacific and Indian oceans pigeons have adapted well to life on islands. They have evolved into many distinctive species, and several frequently fly long distances between the islands in their range. The white and black Torresian imperial pigeon *Ducula spillorrhoa* migrates from New Guinea to northern Australia in August and breeds in huge colonies, mostly on the offshore islands of the Great Barrier Reef. While there, flocks of several thousand birds fly to the rainforests of the mainland every day to feed; in March they return to New Guinea.

In Asia and Africa, the aptly named green pigeons (genus *Treron*) replace the fruit doves and imperial pigeons as the arboreal fruit-eating species in the tropical forests. They lack the specialized gut, however, and grind up the seeds of the fruits they eat. Elsewhere, in Eurasia and the Americas, the various pigeons are less specialized and less spectacularly plumaged. Many pigeons are gregarious and form small to large flocks. Flocks of up to 100,000 woodpigeons *Columba palumbus* have been recorded in Germany, and the eared dove *Zenaida auriculata* of South America breeds in huge colonies of tens of thousands of birds. The flock pigeon *Phaps histrionica* of semi-arid north and central Australia occasionally irrupts in huge flocks of thousands of birds; nineteenth-century explorers described the noise of the flocks as deafening, like "the roar of distant thunder".

FRANCIS H. J. CROME

▲ *Nearly turkey-sized, the three species of crowned pigeons, or gouras, are the largest of all pigeons. They are terrestrial and live in lowland rainforest on the island of New Guinea. Hunted widely for food, they now remain common only in remote areas. Males use their glorious crests in bowing displays during courtship. The Victoria crowned pigeon, shown here, is the smallest of the three and inhabits the northern part of the island.*

ledges. One or two plain white eggs are laid, and the chicks are cared for by both sexes; they leave the nest in 7 to 28 days, depending on the species.

The common street pigeon or rock dove *Columba livia* has adapted well to agriculture and towns. It originally occurred in Eurasia and North Africa, where it nested colonially on cliffs, but it has easily made the transition from cliffs to buildings and now occurs in almost all the world's cities. But this rather drab species gives no indication of the diversity and brilliance of plumage in the family. For instance, the snow pigeon *C. leuconota* from the high plateaus of the Himalaya mountains is a striking white, black, and gray, and the Seychelles blue pigeon *Alectroenas pulcherrima* is deep metallic blue with silver-gray foreparts and a red head with naked wattles.

But perhaps the most beautiful are the 47 species

DRIVEN TO EXTINCTION

Nothing symbolizes human treatment of wildlife and the need for conservation better than the tragic extermination of the dodo *Raphus cucullatus*. It was discovered in 1507 and exterminated by 1680. The dodo of Mauritius was one of three species of massive, flightless, highly aberrant birds on the remote Mascarene Islands, east of Madagascar in the Indian Ocean. Presumably, they derived from pigeon-like ancestors that flew to the islands.

The strange dodo was ash gray with a reddish tinge to its black bill and weighed about 23 kilograms (50 pounds). Its wings were reduced to useless stubs and it had no defense against, or means of escape from, the seafarers who killed it for food, sport, and because they thought it abominably ugly. The pigs, cats, rats, and monkeys that were introduced to the islands may have contributed to its extinction, but basically it fell victim to human persecution.

The very similar white solitaire *R. solitarius* of neighboring Réunion Island was wiped out in the same way by about 1750, but the Rodriguez solitaire *Pezophaps solitaria* managed to survive until perhaps 1800. There are few accounts of the behavior and biology of these species and, indeed, few specimens in museums. They supposedly laid one egg each year, were vegetarian, used their stubby wings for fighting, and were agile runners despite their size and gross proportions.

Like the dodo, the North American passenger pigeon *Ectopistes migratorius* was hunted to extinction. But unlike the dodo, the passenger pigeon was found over much of a continent and was incredibly abundant, possibly the most numerous bird in the world. When white people first came to North America there may have been three to five thousand million or more of the species.

Passenger pigeons underwent irregular migrations within their huge range and were not abundant every year. In good years, however, flocks reached staggering proportions. In 1871 one flock seen over Wisconsin occupied 2,000 square kilometers (850 square miles) and contained 136 million birds; in 1810 a single flock of two and a quarter *billion* birds was seen in Kentucky; and in Ontario a flight of birds moving north from the United States in 1866 was 480 kilometers (300 miles) long and 1.6 kilometers (1 mile) wide and continued for 14 hours. Possibly there were three billion birds in it.

Forest clearing obviously hastened the decline of this species, but the passenger pigeon was exterminated by relentless slaughter just as the bison almost was. Birds were shot, trapped, and poisoned in millions; at one nesting colony in Michigan alone 25,000 birds were killed daily for market during the breeding season of 1874. Over 700,000 a month! Even the commonest bird in the world could not sustain such obscene carnage indefinitely. By the 1880s the species was close to extinction, and the last passenger pigeon died in

Cincinnati Zoo on September 1, 1914. Thus passed one of the greatest wildlife spectacles witnessed by modern man.

The dodo, the solitaire, and the passenger pigeon have not been the only species in this order to suffer extinction. The Mauritius blue pigeon *Alectroenas nitidissima,* Norfolk Island dove *Gallicolumba norfolciensis,* tanna ground dove *G. ferruginea,* bonin wood pigeon *Columba versicolor,* and the silver-banded black pigeon *C. jouyi* have all been exterminated. The small, beautiful, Solomon Islands crowned pigeon *Microgoura meeki* is probably extinct, and many other species, perhaps all those on the small islands of the Pacific where forests are being cleared, are endangered. For such an inoffensive group of birds, pigeons have suffered badly at the hands of humans.

◀▼ *Two notorious extinctions: the passenger pigeon (left) once migrated across North America in hordes that darkened the skies, but the last one died in the Cincinnati Zoo in 1914; the dodo (below) was confined to the island of Mauritius in the Indian Ocean, but was exterminated by early explorers before 1680.*

PARROTS

KEY FACTS

ORDER PSITTACIFORMES
- 2 families • 78 genera
- 330 species

SMALLEST & LARGEST

Bull-faced pygmy parrot
Micropsitta pusio & allied species
Total length: 85 mm (3⅜ in)
Weight: 10–15 g (½ oz)

Hyacinth macaw *Anodorhynchus hyacinthinus*
Total length: 1 m (3¼ ft)
Weight: 1.3 kg (3 lb)

CONSERVATION WATCH
!!! The 9 critically endangered
species are: Puerto Rican amazon
Amazona vittata; Lear's macaw
Anodorhynchus leari; Philippine
cockatoo *Cacatua haematuropygia*;
Spix's macaw *Cyanopsitta spixii*;
Norfolk Island parakeet
Cyanoramphus cookii; night parrot
Pezoporus occidentalis; Fuertes's
parrot *Hapalopsittaca fuertesi*;
yellow-eared parrot *Ognorhynchus
icterotis*; Mauritius parakeet
Psittacula echo.
!! 28 species are listed as
endangered.
! 52 species are listed as
vulnerable.
■ The kakapo *Strigops habroptilus*
is extinct in the wild.

Probably no group of birds is more widely known to the general public than the parrots. Indeed, one species—the budgerigar from inland Australia—rivals goldfish as the most popular pet animal in the world. The popularity of keeping parrots as pets dates from early recorded history: rose-ringed parakeets were known to the ancient Egyptians, and it was probably Alexander the Great who introduced tame parrots from the Far East to Europe. Today, the international trade in live parrots has reached alarming proportions, and there is virtually no city or town without a pet shop selling budgerigars, cockatiels, or lovebirds.

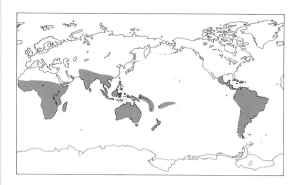

▼ *Two Australasian parrots: the black-capped lory (left) is one of the most conspicuous parrots in New Guinea, and the eastern rosella (right) is common in open eucalypt woodland over much of eastern Australia.*

BRIGHT BIRDS IN BOLD PLUMAGE

Parrots belong to a very distinct order of ancient lineage and are strongly differentiated from other groups of birds. Some distinguishing features are obvious to even a casual observer; most prominent is the short blunt bill with a downcurved upper mandible fitting neatly over a broad, upcurved lower mandible. This unique design enables parrots to crush the seeds and nuts that constitute the diet of most species.

Another conspicuous characteristic is the typical parrot foot, with two toes pointing forward and two turned backward. Parrots show remarkable dexterity, using their feet for climbing or for holding food up to the bill. The skull is broad and relatively large, with a spacious brain cavity. The

extremely muscular tongue is thick and prehensile, and in lorikeets of the subfamily Loriinae it is tipped with elongated papillae for harvesting pollen and nectar from blossoms.

Parrots are renowned for the generally brilliant coloration of their plumage. There are plain or dull-colored parrots, such as the two *Coracopsis* species from Madagascar, but these are few. Green predominates in most species, and is effective as camouflage amidst foliage in the rainforest canopy where many species live. Bold markings, mainly of red, yellow, and blue, are prevalent on the head or wings, and many species have brightly-colored rumps. Unusual plumage patterns are present in the spectacular *Anodorhynchus* and *Cyanopsitta* macaws from South America and some *Vini* lories from the Pacific Islands, all of which are entirely or almost entirely blue, while bright yellow predominates in the plumage of the golden conure *Aratinga guarouba* from Brazil, and the regent parrot *Polytelis anthopeplus* and yellow rosella *Platycercus flaveolus* from Australia. Some *Ara* macaws from Central and South America are almost entirely red, as are some lories from the Indonesian archipelago.

The sexes generally are alike, though females may be appreciably duller, but the eclectus parrot *Eclectus roratus* of Australia is notable in that the bright green males are strikingly different from the predominantly red females.

VARIATIONS ON A THEME
Despite the homogeneity of their basic features, parrots come in all shapes and sizes. Tails may be long and pointed, as in the long-tailed parakeet

Psittacula longicauda from Malaysia and the princess parrot *Polytelis alexandrae* from Australia, or short and squarish as in the short-tailed parrot *Graydidascalus brachyurus* and some *Touit* parrotlets from South America, or there may be ornate feathers as in the Papuan lory *Charmosyna papou* from New Guinea or the *Prioniturus* racket-tailed parrots from Indonesia and the Philippines. Wings can be narrow and pointed, as in the swift parrot *Lathamus discolor* and the cockatiel *Nymphicus hollandicus* from Australia, or broad and rounded, as in the *Amazona* parrots from South America. Prominent erectile head-crests distinguish the cockatoos, which are placed in the separate family Cacatuidae. Other parrots may have elongated feathers on their crowns or hindnecks.

Even in the characteristic bill there are variations in shape, which represent modifications for different feeding habits. A curved, less elongated upper mandible enables the slender-billed corella *Cacatua tenuirostris* of Australia to dig up roots and corms, while similarly-shaped bills of the slender-billed conure *Enicognathus leptorhynchus* of South America and the red-capped parrot *Purpureicephalus spurius* of Australia seem to be ideal for extracting seeds from large nuts. Parrots that feed extensively on pollen, nectar, or soft fruits tend to have narrow, protruding bills.

DISTRIBUTION AND HABITATS
Parrots live mainly in the Southern Hemisphere, and are most prevalent in tropical regions. Once the Carolina parakeet *Conuropsis carolinensis* of North America became extinct in the early part of this century, the northernmost species became the

▲ *A blue-and-yellow macaw. These splendid birds were once common in forests across much of South America, but they have been much reduced because of illegal trafficking for the cage bird trade. They prefer tall palms growing along watercourses. It seems probable that they mate for life, and even when congregating in large flocks the pairs remain in close contact.*

CYG/The Photo Library

▲ Brilliantly plumaged rainbow lorikeets are commonly seen feeding in suburban gardens throughout their range in northern and eastern Australia. The species also occurs widely in the New Guinea region and on many islands in the southwestern Pacific. Strongly gregarious, rainbow lorikeets forage in parties of up to 50 or so, but at night they often congregate in thousands to roost.

▶ The blossom-headed parakeet inhabits hill forest from Bengal to Indochina. In places it remains quite common, and is usually encountered in small flocks or family parties that hurtle through the forest with remarkable speed and agility. It roosts communally and feeds on seeds, nuts and flowers.

slaty-headed parakeet *Psittacula himalayana*, in eastern Afghanistan. The most southerly parrot is the Austral conure *Enicognathus ferrugineus*, which reaches Tierra del Fuego. The strongest representation of parrot species is in Australasia and South America, with 52 species recorded from Australia and 71 from Brazil, but in South America there is a marked uniformity of types. There are parrots in Asia, mainly on the Indian subcontinent, and in Africa, but representation in these regions is surprisingly low. The most widely distributed species is the rose-ringed parakeet *Psittacula krameri*, which occurs in Asia and northern Africa and has been introduced to parts of the Middle East and Southeast Asia. Most of the species with restricted ranges are confined to quite small islands. With a total area of some 21 square kilometers (8 square miles), the Antipodes Islands south of New Zealand are inhabited by two *Cyanoramphus* parrots; one of these, the Antipodes green parakeet *C. unicolor*, is endemic.

Parrots are particularly plentiful in lowland tropical rainforest, although in Australia and parts of South America open country is preferred by many species. While some species, especially those

C.A. Henley

restricted to rainforest, show little capacity to withstand interference with their habitat, others have adapted remarkably well to the human impact and are commonly seen in parks, gardens, or even trees lining city streets. In Australia the galah *Eolophus roseicapillus* is plentiful in many towns and cities, while in downtown São Paulo, Brazil's largest city, small flocks of plain parakeets *Brotogeris tirica* can be seen in parks surrounded by towering buildings.

Parrots tend to be less common at higher altitudes, and species that occur there normally are absent from or are rare in neighboring lowlands.

Distinctive highland forms include the Johnstone's lorikeet *Trichoglossus johnstoniae* in the Philippines, the Derbyan parakeet *Psittacula derbiana* in Tibet, and the yellow-faced parrot *Poicephalus flavifrons* in Ethiopia. Possibly the most interesting of highland parrots is the kea *Nestor notabilis* from the Southern Alps of New Zealand; it is a species that has been much maligned because of its alleged sheep-killing habits.

FAMILIAR AND UNFAMILIAR SPECIES

Parrots are difficult to observe in the wild. Most are predominantly green and live in the rainforest canopy, so sightings usually are little more than momentary glimpses of screeching flocks flashing overhead. Species that inhabit open country or are plentiful near human habitation are conspicuous, and there is much more information on their habits. We know a great deal, for example, about the habits of species such as the eastern rosella *Platycercus eximius* and red-rumped parrot *Psephotus haematonotus* in Australia, Meyers parrot *Poicephalus meyeri* in Africa, and the monk parakeet *Myiopsitta monachus* in South America, but virtually nothing about many forest-dwelling species in New Guinea, Central Africa, and South America.

The Australian night parrot *Pezoporus occidentalis* is nocturnal, and there are reports of some normally diurnal species being active on moonlit nights. Migration of swift parrots and *Neophema* species across Bass Strait, in southern Australia, usually takes place at night. In most parrots, patterns of daily activity are typical of birds in tropical regions: peak periods in the morning and toward evening, and low activity during the heat of the day.

Many parrots are gregarious. Pairs and family parties come together to form flocks, which in arid areas may build up to enormous sizes after breeding has been brought on by favorable conditions. At such times, massive flocks of little corellas *Cacatua saguinea* and budgerigars darken the skies of inland Australia.

◄ The hanging-parrots, so called because of their extraordinary habit of roosting, bat-like, upside down, constitute a group of ten species best represented in Indonesia. They seem closely related to the lovebirds of Africa and, like them, transport material for their nests among the feathers of the rump. This is the blue-crowned hanging-parrot, which is widespread in Malaysia, Sumatra, and Borneo.

Morten Strange

▶ Pesquet's parrot (left) inhabits hill forest in New Guinea, where it is widespread but rare. It feeds on soft fruit and nectar, and the largely naked head is probably an adaptation to avoid having plumage matted with sticky juice and nectar. The hawk-headed parrot (right) of Amazonia has a striking ruff of long, colorful, erectile feathers on the nape. It is a noisy, conspicuous parrot that is usually encountered in small parties.

The flight of most parrots, especially the smaller ones, is swift and direct. Some have a characteristically undulating flight produced by wing beats being interspersed with brief periods of gliding. In the larger species it is variable; macaws are fairly fast fliers and their wing beats are shallow, but the buoyant flight of *Probosciger* and *Calyptorhynchus* cockatoos is conspicuously slow and labored. The kakapo *Strigops habroptilus* is the only flightless parrot.

The distinctly metallic call-notes of most parrots are harsh and unmelodic, generally being based on a simple syllable or combination of syllables. Variation comes primarily from the timing of repetition. Larger species normally have raucous, low-pitched calls, while small parrots give high-pitched notes. The mimicry of captive parrots is well known, so it is surprising that there are very few convincing reports of wild birds imitating other species.

FOOD AND FEEDING

Most parrots eat seeds and fruits foraged from treetops or on the ground. Lories and lorikeets of the subfamily Loriinae are strictly arboreal, and feed on pollen, nectar, and soft fruits. Insects are often found in crop and stomach contents; wood-boring larvae are an important food item for some of the black cockatoos from Australia. Mystery still surrounds the diet of *Micropsitta* pygmy parrots, which seem to take lichen from the trunks and branches of trees, but at times they have been observed foraging for termites.

When feeding, a parrot makes full use of its hooked bill; while climbing among foliage it often uses the bill to grasp a branch onto which it then steps. Many species use a foot as a "hand" to hold food up to the bill, and with the bill they expertly extract kernels from seeds and discard the husks.

BREEDING BEHAVIOR

The age at which parrots reach sexual maturity varies, but in general it is three or four years in larger species and one or two years in small birds. As far as can be ascertained from observations, most species are monogamous and the majority remain paired for long periods, perhaps for life. Notable exceptions are the kea, which is

polygamous, and the kakapo, a lek-display species with males almost certainly taking no part in incubation or care of the young. Pairs and family groups are usually discernible within flocks. Courtship displays are simple, with bowing, wing-drooping, wing-flicking, tail-wagging, foot-raising, and dilation of the eye pupils being the more common actions.

Parrots usually nest in hollows in trees or holes excavated in termite mounds, occasionally in holes in banks, or in crevices among rocks and cliff-faces. The ground and night parrots from Australia and some populations of *Cyanoramphus* parakeets on New Zealand islands nest on the ground, usually under or in grass tussocks. Monk parrots from South America gather twigs to build a huge communal nest in a tree, and each pair has its own breeding chamber.

Eggs are normally laid every other day, and clutches vary from two to four or five, sometimes up to eight for small parrots. Incubation starts with or immediately after the laying of the second egg, but there is mounting evidence that this can vary

Leo Meier/Weldon Trannies

◄ *Few sights are more evocative of the arid interior of Australia than a tree festooned with little corellas, or a massed flock of these birds rising from the ground in a roar of wings and a cacophony of harsh screeches. They are intensely gregarious, and flocks may sometimes number thousands of birds. They roost communally, invariably near waterholes. This is a habit that was exploited by many of the early desert explorers, who followed the gathering flocks at dusk to be led to water.*

individually. Generally the female alone incubates. The duration of incubation for small parrots is from 17 to 23 days, but for the large macaws it can be up to five weeks. Newly-hatched chicks are blind and naked or have sparse dorsal down, which in most species is white.

Young parrots develop slowly, and remain in the nest for three to four weeks in the case of the smallest species, and up to three or four months for the large macaws. After leaving the nest, young birds are fed by their parents for a brief time while learning to fend for themselves; young black cockatoos are fed by their parents for up to four months after leaving the nest.

Juveniles generally resemble females or are duller than either adult sex. There are species, such as the crimson rosella *Platycercus elegans* from Australia and some *Psittacula* species, that have a distinct juvenile plumage. A striking difference between adults and juveniles occurs in the vulturine parrot *Gypopsitta vulturina* from Brazil: in adults the bare head is sparsely covered with inconspicuous "bristles", but in juveniles the head is well covered with pale green feathers. The time taken for juveniles to attain adult plumage varies greatly between species; it may be within months of leaving the nest, or up to three or four years.

CONSERVATION OF PARROTS

Ten extinct species of parrot are represented by specimens in museums, while others are known from subfossil material or reports in the writings of early explorers. Probably the best known of these extinct species is the Carolina parakeet; the last living bird died in the Cincinnati Zoo on February 21, 1918. Even for this species the causes of extinction will never be fully understood, but the loss is a warning that should be heeded if parrots are to be protected from the serious threats they now face in virtually all parts of their range. Habitat destruction is by far the most serious threat, especially the clearing of tropical forest. Of special concern are parrots confined to small islands, where the habitat is finite and cannot be extended.

Parrots are among the world's most endangered birds, in part because of the live-bird trade. Methods of capture are wasteful and often inhumane, and the levels of trapping are having severe impacts on populations already suffering from loss of habitat.

JOSEPH FORSHAW

THE PARROT THAT CANNOT FLY

In just about every aspect of its biology, the kakapo or owl parrot *Strigops habroptilus* of New Zealand is unique. It is much heavier than other parrots. It is nocturnal. And although its wings are well developed, there is no sternal keel for attachment of the wing muscles, and hence the bird cannot fly! Its method of feeding is peculiar, and produces telltale signs of its presence in any given area. The parrots chew the leaves or stems, extract the juices, and leave behind, hanging on the plant, tight balls of macerated fibrous material, which are then bleached white by the sun. But its strange breeding behavior sets the kakapo apart from all other parrots most decisively. Male courtship is a lek-display involving social displaying and "booming" from inflated gular air-sacs at excavated bowl-like depressions or "courts" on arenas or traditional display grounds. Females respond to the booming and come to the courts, where mating takes place. Males take no part in nesting, leaving the female to excavate a burrow under rocks or tree roots, in which she lays one or two, rarely three, eggs, and where she alone rears the chicks. Tragically, the kakapo is now seriously endangered. A major conservation effort by New Zealand wildlife authorities is underway, with funding coming from both government and corporate sources.

Don Merton

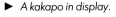 *A kakapo in display.*

TURACOS & CUCKOOS

KEY FACTS

ORDER CUCULIFORMES
• 2 families • *c.* 40 genera
• *c.* 150 species

SMALLEST & LARGEST

Little bronze-cuckoo
Chrysococcyx malayanus
Total length: 14–15 cm (5½–6 in)
Weight: about 30 g (1 oz)

Great blue turaco *Corythaeola
cristata*
Total length: about 90 cm (3 ft)
Weight: 1–1.2 kg (2¼–2⅔ lb)

CONSERVATION WATCH
!!! The black-hooded coucal
Centropus steerii is critically
endangered.
!! The Prince Ruspoli's turaco
Tauraco ruspolii, green-billed
coucal *Centropus chlororhynchus*,
and banded ground-cuckoo
Neomorphus radiolosus are listed
as endangered.
! There are 6 vulnerable species:
Bannerman's turaco *Tauraco
bannermani*; Sunda ground-
cuckoo *Carpococcyx radiceus*;
Sunda coucal *Centropus nigrorufus*;
Cocos cuckoo *Coccyzus ferrugineus*;
rufous-breasted cuckoo *Hyetornis
rufigularis*; red-faced malkoha
Phaenicophaeus pyrrhocephalus.

T he order Cuculiformes consists of two quite different families—the turacos (louries or plantain-eaters) of the family Musophagidae, and the cuckoos of the family Cuculidae—which are united chiefly because their egg-white proteins are similar. Some ornithologists classify them in two separate orders because of their many anatomical differences, totally different juvenile development, different patterns of molt, and the different lice living among their feathers. Both groups have a long evolutionary history, being known as fossils from about 40 million years ago. At present, the turacos are confined to Africa south of the Sahara. The cuckoos are virtually cosmopolitan but are best represented in the tropics and subtropics.

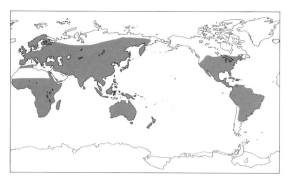

simple frail platforms of sticks. The young hatch at an advanced stage of development, and have thick down and open eyes (or nearly so). They are fed by regurgitation, and usually scramble out of the nest long before they can fly.

TURACOS

Turacos are rather long-necked birds with long tails, short rounded wings, and erectile, laterally compressed crests, except in one species. The great blue turaco is the largest, but all the other species are roughly the same size, about 45 centimeters (17¾ inches) in length and weighing 250 to 350 grams (9 to 12½ ounces). Their feet are semi-zygodactylous (the fourth or outer toe is at right angles to the main axis of the foot and is capable of being directed backwards or forwards). They are gregarious, noisy birds, going about in parties of five to ten in forest and savanna, flying rather weakly from tree to tree with gliding and flapping flight, but running and bounding nimbly among the branches within a tree, communicating with harsh loud barking calls that can be heard from afar and may be recognized by other creatures such as antelopes as signals of alarm. In general they are sedentary, or at least non-migratory. Five species are dull-plumaged, grayish, brownish or whitish birds that inhabit the savannas—for example, the gray go-away bird *Corythaixoides concolor*. The rest are brightly colored in greens, reds, purples, and blues, which are formed by special pigments such as turacin, apparently unique in the animal kingdom. These brightly colored species typically live in forests. Turacos eat mostly fruit, leaves, buds, and flowers but also take insects, especially when breeding. Their nests, always in trees, are

▲ *The white-cheeked turaco has a
restricted distribution in Ethiopia and
Somalia in northeastern Africa. It
frequents dense brush and forest edges,
feeding largely on fruit.*

125

The family is subdivided into six groups of great diversity, classified more on breeding habits and geographical distribution than on structural characters, though all have zygodactylous feet (two toes pointing forwards and two backwards).

True cuckoos Members of the subfamily Cuculinae have parasitic habits like those of the European bird and are confined to the Old World. Mostly they are drab, black and white, or black birds, differing more or less between the sexes, with long pointed wings. The smaller bronze-cuckoos of the tropics and Southern Hemisphere may, however, be brightly colored—for example, the emerald cuckoo *Chrysococcyx cupreus* of Africa, which is green and gold. Cuckoos generally seem to be solitary because the males are often very conspicuous and noisy with persistent bouts of loud, striking, and rather monotonous calls, even at night, whereas the females are unobtrusive and have different, even muted, calls. Probably few species, if any, form simple pairs for breeding; the male's territory may overlap with those of more than one female, and they may therefore be promiscuous. Species are to be found in a variety of habitats, from open moorland in northwestern Europe to tropical rainforest. Most species (at least, outside the tropics) are strongly migratory. Hairy caterpillars form a large part of their diet.

In spite of all sorts of claims, these cuckoos do lay their eggs directly into the nests of other birds, however improbable or impossible this may seem when the nest that they parasitize is small and enclosed. The female cuckoo usually removes an egg of the host when laying her own; her eggs generally hatch earlier, and the young cuckoo evicts unhatched eggs or young of the host within three to four days of hatching. However, some large species, such as the channel-billed cuckoo *Scythrops novaehollandiae,* often lay several eggs in the host's nest; rather than evicting the host's eggs or young, the young cuckoos out-compete them for food and they die. Cuckoos' eggs often mimic those of the host, especially if the host makes an open cup-shaped nest, and different females within one species of cuckoo may each be adapted to parasitize one species of host.

Nest-building cuckoos The subfamily Phaenicophaeinae comprises the malkohas of tropical Asia, quite large, long-tailed birds with bare, brightly colored faces; large, lizard-eating cuckoos (genus *Saurothera*) and other diverse species of the Caribbean; and the Coccyzus cuckoos of the New World.

Anis and guiras The anis and guiras of the subfamily Crotophaginae are a small New World group of gregarious, short-winged, long-tailed cuckoos that breed communally and maintain group-territories. Characteristically their bills are deep and laterally compressed.

Road-runners and ground cuckoos The members of the subfamily Neomorphinae of the

Richard T. Mills

▲ *Cuckoos are notorious for their habit of laying their eggs in the nests of other birds, to be raised by the unwitting fosterers. However, it is not true that all cuckoos do this, nor are cuckoos the only birds that practice this strategy. The European cuckoo lays its eggs in the nests of a wide range of hosts, including warblers, wrens and pipits. This is a juvenile bird.*

CUCKOOS

For many people, the name "cuckoo" will conjure up a symbol of summer—a rather sleek, long-tailed grayish bird with long pointed wings, a well-known call, and the habit of laying eggs in the nests of other species, from which its chicks evict the eggs or young of the host. That image is based on the performance of one species, the European cuckoo *Cuculus canorus,* around which a huge body of myth and speculation has been built up over the centuries. In fact, of the 130 or so species in the family Cuculidae, only about 50 are truly parasitic.

Americas, with one genus (*Carpococcyx*) in Asia, are terrestrial birds, some of which rarely fly and some of which inhabit arid regions. The striped cuckoo *Tapera naevia*, pheasant cuckoo *Dromococcyx phasianellus*, and pavonine cuckoo *D. pavoninus* are parasitic like the true cuckoos but are classified here because they appear to be so like road-runners.

Couas The subfamily Couinae comprises 10 non-parasitic species confined to Madagascar; one species may now be extinct. They are long-legged birds, the size of pigeons. Some are brightly colored and have naked skin on the head.

Coucals The coucals of the subfamily Centropodinae are terrestrial cuckoos with long straight hind-claws (thus sometimes called "lark-heeled cuckoos"), which build domed nests, lay white eggs, and are confined to the Old World. Their persistent hooting monotonous calls—praying for rain or giving thanks for fine weather, according to local folklore in Africa—proclaim their presence, although most species tend to skulk in thick undergrowth in woodlands.

S. MARCHANT

R. Drummond

▲ The pheasant coucal is common in rank grasslands in New Guinea and in northern and eastern Australia.

▲▶ Two aberrant cuckoos. The channel-billed cuckoo (above) is unusual among cuckoos in that it feeds largely on fruit; it breeds in northern and eastern Australia and spends the winter in Indonesia and New Guinea. The original inspiration for the famous cartoon character, the roadrunner (right) inhabits deserts of the American southwest. As its name suggests it is terrestrial, and feeds on lizards, snakes, and other small animals.

127

KEY FACTS

ORDER STRIGIFORMES
• 2 families • 24 genera
• c. 162 species

SMALLEST & LARGEST

Least pygmy owl *Glaucidium minutissimum*
Total length: 12–14 cm (4¾–5½ in)
Weight: less than 50 g (1¾ oz)

Eurasian eagle owl (European race) *Bubo bubo*
Total length: 66–71 cm (26–28 in)
Weight: 1.6–4 kg (3½–8¾ lb)

CONSERVATION WATCH
!!! There are 4 critically endangered species: forest owlet *Athene blewitti*; Anjouan scops owl *Otus capnodes*; Seychelles scops owl *Otus insularis*; Grand Comoro scops owl *Otus paulani*.
!! 5 species are listed as endangered: Madagascar red owl *Tyto soumagnei*; Philippine eagle-owl *Bubo philippensis*; Blakiston's fish owl *Ketupa blakistoni*; lesser eagle-owl *Mimizuku gurneyi*; rufous fishing owl *Scotopelia ussheri*.
! 17 species are vulnerable.

OWLS, FROGMOUTHS & NIGHTJARS

The Strigiformes (owls) and Caprimulgiformes (frogmouths, nightjars, and their allies), are both well-defined groups, and even for people with little ornithological training the members of each are instantly recognizable. The two orders share many characteristics and are thought to be distantly related. Both are crepuscular (twilight-active) and nocturnal (night-active). Their soft plumage is typically in "dead-leaf" and "mottled-bark" colors and patterns, which are most refined in nightjars and frogmouths. Immobility and posture add to the effectiveness of their camouflage. With flattened feathers, bill tilted skyward, and eyes closed to a slit, a disturbed frogmouth is indistinguishable from the broken branch of a tree. So cryptic are the nightjars as they crouch, roosting on the ground, that photographs of them become "find-the-hidden-bird" puzzles. The owls also flatten their plumage when slightly disturbed but with their longer legs and wider eyes they are more obvious; they roost in more hidden places and flush more readily than many of the caprimulgiforms.

OWL DISTRIBUTION

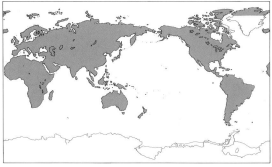

▼ *Many night-birds, like this common potoo in South America, can be located by their eye-shine reflecting back from a powerful torch.*

ADAPTED TO DIM LIGHT

These birds are more often heard than seen; their distinctive calls, described as startling, strange, or weirdly beautiful, often carry across the countryside. Their calls and mysterious nocturnal habits have been the basis for much superstition, from shrieking ghosts to the ancient belief that the nightjars steal milk from goats, hence one of their common names, goatsucker. In fact, they flit around goats and other livestock in pursuit of the insects attracted to them.

Life in dim light has led to some remarkable sensory adaptations: large eyes with good vision in poor light; and, in total darkness, navigation by echolocation (oilbirds) and hunting by exceptional hearing (barn owls).

They all have rather large heads. Most species have large eyes: forward-facing in the owls for increased binocular vision; more laterally placed in the caprimulgiforms. Their eyes are specialized for vision in poor light, with more rods (light-sensitive elements) than diurnal (day-active) birds. Nevertheless, most species appear to need some light before they are able to hunt. While most species habitually hunt in poor light, they can see well by day and some occasionally hunt in daylight (for example, the barn owl *Tyto alba*, the burrowing owl *Athene cunnicularia*, and the barking owl *Ninox connivens*); the northern hawk-owl *Surnia ulula* is largely diurnal.

The remarkable barn owls (genus *Tyto*), with rather small eyes, have the most exceptional hearing. They are able to catch prey in total

M.P.L. Fogden/Bruce Coleman Ltd

Richard Cannings

◀ In many owls, the young leave the nest well before they can fly; the parents continue to feed them for weeks, and in a few cases for months, thereafter. These young saw-whet owls have already donned the distinctive juvenile plumage but have still not reached independence. So-called because one of its calls resembles the sounds produced in sharpening a saw, the saw-whet owl inhabits dense coniferous forests in North America.

OWL SKULL

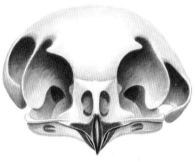

▲ In some barn owls the ears are asymmetrical in both size and shape. This enhances the "stereo" effect—the subtle difference between the sound signal reaching one ear relative to that reaching the other ear—and enables very precise location of prey.

darkness, guided by sound alone. They and several other owls have facial masks to catch sound, and some have asymmetrical ear openings. Either the external feathering or the skull itself is modified so that sounds reach one ear at a slightly different time to the other; by turning its head the owl can locate the source of a low sound, such as a mouse chewing grain, very precisely. Oilbirds also have a remarkable adaptation for night navigation. They nest and roost gregariously, deep in caves. At night a mass of birds navigates through the cave by making audible (to humans) clicks and using the echoes that return to their ears to guide them from the cave; once outside they cease clicking.

Most species have soft, loose plumage, with frayed trailing edges to the flight feathers of their wings and tail, for noiseless hunting flight. Exceptions are the fishing owls (genus *Scotopelia*) of Africa and the oilbirds, which are hunters of fish and gatherers of fruit, respectively, presumably with little need for silent flight; both have firmer feathers.

The two groups differ most obviously in their bill and feet. The owls have a sharp, hooked bill and strong legs and feet, with sharp curved talons for their predatory lifestyle. The nightjars and their allies have a broad flattened bill, an enormous gape, and small, weak feet and legs. Both orders have reversible outer toes and can perch with two toes forward, two back. The barn owls and the caprimulgiforms have a serrated edge on the talon of their middle toe, perhaps as an aid to grooming.

OWLS

Currently, the owls are split into two families. All have rather long, broad wings.

Barn owls

Members of the family Tytonidae are medium-sized owls with heart-shaped faces, inner toes as long as their middle toes, and long bare legs. The bay owls (genus *Phodilius*) are currently placed in this family but may resemble barn owls (genus *Tyto*) only superficially.

Hawk-owls or true owls

The family Strigidae are small to large owls with rounded heads, large eyes, stout, sometimes feathered legs, and the inner toe shorter than the middle toe. Some show little sign of a mask; others are partially masked; and some have a full, rounded mask. Several species have two tufts of erectile feathers or "ears" which they can raise in emotion and which may help with concealment by disguising the owl's outline. One species, the maned owl *Jubula lettii* of west Africa and the Congo, has voluminous crown and nape feathers.

The female of most owl species is larger than the male, sometimes considerably so: female Tasmanian masked owls *Tyto novaehollandiae* weigh an average of 965 grams (34 ounces), males a mere 525 grams (18½ ounces). But, in some of the *Ninox* species, it is the male that is larger: for example, the female barking owl weighs 510 grams (18 ounces), the male 680 grams (24 ounces).

KEY FACTS

ORDER CAPRIMULGIFORMES
• 5 families • *c.* 24 genera
• *c.* 102 species

SMALLEST & LARGEST

Donaldson-Smith's nightjar
Caprimulgus donaldsoni
Total length: 19 cm (7½ in)
Weight: 21–36 g (¾–1¼ oz)

Papuan frogmouth *Podargus papuensis*
Total length: 50–60 cm (20–23½ in)
Weight: 300–570 g (10½–20 oz)

CONSERVATION WATCH

!!! There are 3 critically endangered species: white-winged nightjar *Caprimulgus candicans;* Puerto Rican nightjar *Caprimulgus noctitherus;* Jamaican pauraque *Siphonorhis americanus.*
!! The New Caledonian owlet-nightjar *Aegotheles savesi* is listed as endangered.
! There are 3 vulnerable species: Vaurie's nightjar *Caprimulgus centralasicus;* Itombwe nightjar *Caprimulgus prigoginei;* satanic nightjar *Eurostopodus diabolicus.*

NIGHTJARS AND THEIR ALLIES

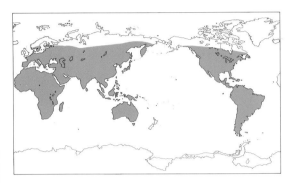

The Caprimulgiformes are divided into five families.

Oilbird
The oilbird *Steatornis caripensis* is the sole member of the family Steatornithidae. It has a fan-like tail, long broad wings, and is dark brown with white spots and black bars. Adult size is about 30 centimeters (12 inches). In common with the other caprimulgiforms, it has a strong hook-tipped bill, a wide gape surrounded by bristles, and large eyes.

Frogmouths
Members of the family Podargidae are the largest of the caprimulgiforms. They have been described as the most grotesque of birds, with a great flat shaggy head dominating the body, which tapers from it. The massive bill, surrounded by large tufts of facial bristles, as wide as it is long and heavily ossified (hardened like bone), acts as a heavy snap-trap. Their legs are short and weak.

Potoos
The potoos (family Nyctibiidae) resemble frogmouths in their arboreal roosting habit and color pattern. Yet their broad, weakly ossified bill, which is surrounded by relatively few bristles, and their aerial hawking behavior, ally them with the nightjars.

Owlet-nightjars
These birds (family Aegothelidae) are somewhere between a nightjar and an owl in appearance, but their closest relatives are the frogmouths. They have a broad flat bill almost hidden by bristles. Their feet are slightly stronger than those of the other caprimulgiforms, and their legs longer, perhaps because they run about more.

Nightjars
The nightjars (family Caprimulgidae) are a large group and comprise about half of the species in the Caprimulgiformes. They have long pointed wings and swift flight, a wide gape, stubby bill, and brightly colored mouth (usually pink), shown in threat. Their legs and feet are weak, much reduced, and rarely used. The nightjars are fairly uniform in appearance, but variations on the basic form include the standard-winged nightjar *Macrodipteryx vexillaria* and pennant-winged nightjar *M. longipennis,* which have extraordinary trailing feathers used in courtship, and the long-tailed nightjar *Caprimulgus climacurus* with a long gradated tail; these three live in Africa. In North America, members of the subfamily Chordeilinae are called nighthawks.

► *Roosting in trees, a frogmouth usually spends the day in a distinctive posture, more or less upright and oriented along the branch; in this position its resemblance to a broken-off stub is extraordinary. But the tawny frogmouth is a sociable bird, and sometimes pairs and even families cuddle up together on a branch.*

Robert Cook/Weldon Trannies

◀ Mostly eyes and fluff, an Australian owlet-nightjar pauses at the entrance to its nest hollow. This bird is common in woodlands across Australia; several other species occur in New Guinea. Owlet-nightjars are unusual among caprimulgids in that they roost and nest in cavities in trees.

Tom & Pam Gardner

DISTRIBUTION AND HABITATS

Owls are cosmopolitan in distribution. They occur on all continents except Antarctica and are absent from some oceanic islands. *Strix* and *Otus* are widespread genera, the latter mostly in tropical areas; they do not occur in the Australia–Papua New Guinea region, where they are replaced by the genus *Ninox*. Some species such as the barn owl and the short-eared owl *Asio flammeus* are among the most widely distributed of all birds. In contrast, the Palau owl *Pyrroglaux podargina* is found only on the Palau islands in the western Pacific Ocean. Habitat destruction and introduced animals have taken their toll and pushed some of these island owls toward extinction.

The majority of owl species inhabit woodlands and forest edges. A few species prefer treeless habitats: for example, the snowy owl *Nyctea scandiaca* of Arctic tundra regions, and the elf owl *Micrathene whitneyi* of the southwestern deserts of the USA. Some, long-legged, species are terrestrial and live in flat grasslands (the grass owl *Tyto capensis* of Africa, and India to Australia) or

marshes (the marsh owl *Asio capensis* of Africa), or among rocks. Various species can be found in most habitats, from tundra and desert to rainforest and swampland, from wilderness to the suburbs; the great horned owl *Bubo virginiatus* occurs in most habitats of North America. Some are more specific in their habitat requirements than others: the northern hawk owl lives in the northern conifer forests (taiga) of North America and Eurasia; the white-throated owl *Otus albogularis* is found in the cloud forests of the Andes; and the Peruvian screech owl *O. roboratus* likes habitat with mesquite and large cacti in arid parts of Peru.

The nightjars and their allies have a similar distribution to owls but are not found at such high latitudes or on as many small islands. Nor are they found on the islands of New Zealand. The oilbird is found only in the neotropics (Guyana to Peru and Ecuador, and Trinidad); it depends on suitable caves, and ranges out from them to forage locally —most are in the mountains, but in Trinidad sea-caves along the rocky coast are used. The potoos prefer open woodland of cultivated areas such as

▶ *Largest of Australian owls, the powerful owl roosts by day in trees, usually selecting places that are fairly open all around but with a screen of foliage above. This one is roosting near its offspring. Feeding largely on possums, the powerful owl frequently holds the remains of its night's meal, tail dangling, clamped in its talons throughout the following day, often polishing it off as a sort of wake-up snack the next evening before going off to hunt again.*

Geoff Longford

coffee plantations in tropical Central and South America. Frogmouths are arboreal inhabitants of forest, woodland and forest edge: *Batrachostomus* species occur from India to Malaysia; and the *Podargus* species in New Guinea, the Solomon Islands, and Australia. The tawny frogmouth *P. strigoides* is found throughout Australia in most habitats, but avoids dense rainforest and treeless desert. Owlet-nightjars live in rainforest or open forest and woodland in New Guinea, Australia, and New Caledonia. The nightjars are widely distributed in warmer parts of the world and have a diversity of habitat preferences. In Africa, for example, there is a different *Caprimulgus* species in almost every habitat.

FEEDING AND BREEDING

Owls drop down from a perch to catch mammals (from mice to hares, depending upon the size of the owl) or insects on the ground. They also snatch insects from foliage, and large species grab arboreal mammals while smaller species hawk insects in the air. They catch prey with their feet and may reach down and dispatch it with a few bites. Small prey is often lifted to the bill with one foot, in the manner

of a parrot; it is swallowed whole. Large prey is held in the feet and dissected with the bill. Once or twice a day owls regurgitate a pellet containing the fur, most bones, chitinous insect remains, and other indigestible parts of their prey.

The caprimulgiforms do not produce a pellet. Oilbirds hover to pluck a variety of fruits and seeds, which they locate by sight and scent. Frogmouths pounce from a perch to catch non-flying animal prey in their massive gape and heavy bill; they batter prey to soften it before swallowing it whole. Owlet-nightjars take some insects and frogs on the ground and dart out to snatch termites, moths and other insects from the air. Most aerial of all are the fast flying nightjars and nighthawks which commonly trawl, open-mouthed, for airborne insects.

Since they are more often heard than seen, it is not surprising that many of these birds are named for the calls they make. All owls call, especially as the breeding season approaches. They have a variety of shrieks, hoots, and barks, which are typical of individual species. Several owls sing quite musically. Some screech-owls (genus *Otus*) duet, the male and female each taking turns to complete their section of a song. Caprimulgiforms also have a variety of far-carrying calls: low drumming in frogmouths; churring in owlet-nightjars; and various other screams and shrieks, hence the name "nightjar". However, some, like the whip-poor-will *Caprimulgus vociferus* of North America, make quite melodious whistles, and one African species, the fiery-necked nightjar *C. pectoralis,* sings "Good Lord, deliver us".

Owls nest in a hole in a tree or cliff, an old building, or the old stick nest of a crow or raptor. The burrowing owl takes over a gopher hole, eagle owls sometimes dig a nest cavity in the side of an anthill, and the snowy owl nests on the open ground in a scrape to which it adds a little lining. A woodpecker hole, drilled in a cactus, is used by the the elf owl. Some caprimulgiforms build a nest. Regurgitated fruit, which sets firm, forms the oilbird's nest, which is built in recesses in the cave and added to each season. Frogmouths build a flimsy nest of sticks on a horizontal branch (*Podargus* species) or use a pad of down from the birds themselves, plus spider webs and lichen (*Batrachostomus* species). Owlet-nightjars nest in a hollow in a tree or occasionally a bank; potoos use a depression on a branch; nightjars and nighthawks nest unceremoniously on bare ground or occasionally on an epiphyte.

Perhaps because they are nocturnal, many species do not seem to have elaborate courtship displays. The male owls feed the females during courtship. During the breeding season, the second primary in each wing of the male standard-winged nightjar projects about 35 centimeters (14 inches) beyond its neighbors and is shown to effect in slow aerial breeding displays. The bird nips them off after the displays cease.

Some owl species lay a similar-sized clutch of eggs each year during a regular breeding season (for example, two to three eggs for *Ninox* species); other vary the start of breeding and the clutch size quite dramatically according to seasonal conditions. The snowy owl lays up to 14 eggs in a year when its lemming prey is abundant, but two to four when prey is scarce. Incubation takes between four and five weeks depending upon the species, and the young must be brooded for a few weeks. The male forages and the female incubates, then both parents feed the young by offering them food. The nestlings often leave the nest to perch nearby when still downy, and in some species the family may stay together for several months.

The caprimulgiforms have a clutch of one to four eggs. Because the risk of predation is high, some ground-nesting nightjars have short incubation times (17 or 18 days in the Northern Hemisphere, but longer in Australia) and the young are semi-precocial—within hours they can totter around. They stay with their parents until migration. Oilbirds feed their nestlings on regurgitated oily fruit for 120 days, until they reach adult size. They have long been collected by South American Indians for their fat, used for cooking and in lamps.

Typically, owls and caprimulgiforms are solitary; the gregarious oilbirds are an exception. In both orders some species are resident, others partially or totally migratory. For example, some European populations of *Otus* owls migrate to Africa for the winter. Most temperate-zone species of nightjar spend winter in the tropics and some tropical species are partial or total migrants. Rather than departing the North American winter, however, each year the common poorwill *Phalaenoptilus nuttallii* hibernates, clinging to the sides of a rock crevice. Its heart rate and respiration drop to almost unmeasurable levels, and its temperature falls from about 41°C (105°F) to about 19°C (66°F). It is one of the very few birds that hibernate regularly.

PENNY OLSEN

▼ The great gray owl inhabits the vast subarctic coniferous forests of the Northern Hemisphere. It is one of the largest of owls, but its bulk is mainly feathers and in fact it weighs less than half as much as many other owls more or less its equal in size. It nests in the abandoned nests of other birds of prey, never adding to them, although it will often rearrange the material.

Jeff Foott/Auscape International

SMALLEST & LARGEST

Bee hummingbird *Mellisuga helenae*
Total length: 6.4 cm (2½ in)
Weight: 2.5 g (¹⁄₁₀ oz)

White-naped swift *Streptoprocne semicollaris*
Total length: 25.4 cm (10 in)
Weight: 185 g (7²⁄₅ oz)

CONSERVATION WATCH
!!! 8 species are listed as critically endangered: Honduran emerald *Amazilia luciae*; turquoise-throated puffleg *Eriocnemis godini*; black-breasted puffleg *Eriocnemis nigrivestis*; Bogotá sunangel *Heliangelus zusii*; scissor-tailed hummingbird *Hylonympha macrocerca*; sapphire-bellied hummingbird *Lepidopyga lilliae*; hook-billed hermit *Ramphodon dohrnii*; Juan Fernández firecrown *Sephanoides fernandensis*.
!! There are 7 species listed as endangered: Esmereldas woodstar *Acestrura berlepschi*; little woodstar *Acestrura bombus*; chestnut-bellied hummingbird *Amazilia castaneiventris*; Táchira emerald *Amazilia distans*; Oaxaca hummingbird *Eupherusa cyanophrys*; white-tailed hummingbird *Eupherusa poliocerca*; short-crested coquette *Lophornis brachylopha*.
! 20 species are vulnerable.

SWIFTS & HUMMINGBIRDS

Swifts, crested swifts, and hummingbirds are generally classified as three families (Apodidae, Hemiprocnidae, and Trochilidae, respectively) in the order Apodiformes. They share some anatomical features, particularly the relative length of the bones of the wing, which is related to their rapid wing beats and flight behavior. The connection between swifts and crested swifts seems clear, but the inclusion of the very dissimilar hummingbirds in this order has often been challenged. Any communality of ancestry is indeed old.

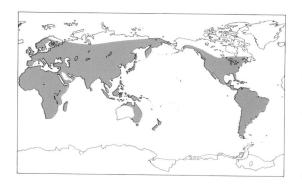

SWIFTS

Swifts, with their narrow swept-back wings, have a well-deserved reputation for being among the fastest flying birds. They range in size from the pygmy swiftlet *Collocalia troglodytes* of the Philippines and pygmy palm swift *Micropanyptila furcata* of Venezuela, which weigh less than 6 grams (¼ ounce), up to the white-naped swift *Streptoprocne semicollaris* of Mexico and purple needletail *Hirundapus celebensis* of the Philippines, both of which approach 200 grams (7 ounces). All are predominantly dark brown or sooty, with some areas of white or gray, and they have short legs with strong claws. All swifts pursue and capture their food, mostly insects, on the wing and stay aloft throughout the day, perching only at their overnight roosts. Sometimes the food ball or bolus taken to a nestling will contain mainly swarming insects such as termites, mayflies or aphids, as well as winged ants, wasps, and bees. At other times up to 60 different kinds of insects and spiders and several hundred individual prey items can be found in a single bolus.

Although most numerous in the tropical areas of the world the 80 or so species of swifts are widely distributed and even occur in Scandinavia, Siberia, and Alaska. The common swift *Apus apus* and alpine swift *A. melba* of Europe, the white-throated needletail *Hirundapus caudacutus* of Siberia and the chimney swift *Chaetura pelagica* of eastern North

▶ *The black swift of western North America nests in rock crevices in cliffs, often behind waterfalls.*

◀ A female Allen's hummingbird feeds her brood of young. As in most hummingbirds, nest-building, incubation, and care of the young is left entirely to the female. The usual clutch is two, and two or three broods are raised each season. This hummingbird is common in coastal California, often nesting in city parks and gardens.

George D. Lepp/Comstock

America all make long migration flights, often over stretches of ocean, to Southern Hemisphere wintering grounds. Even on the breeding grounds some swifts regularly spend the night on the wing.

Many swifts use secretions of their salivary glands in nest building. (Members of the New World subfamily Cypseloidinae do not do this, however, and their nests of mosses, ferns, and other plant material are placed near or behind waterfalls). The salivary glands enlarge during the breeding season to produce a sticky material which, in the genus *Chaetura,* is used to glue together small sticks to form the nest and also to attach it to the vertical wall of a hollow-tree nesting site; while in flight, the birds break dead twigs from the tops of trees. The use of saliva in nest building is most highly developed in some of the smaller cave-inhabiting swifts, known as cave swiftlets, of Southeast Asia, where saliva makes up the bulk of the nest. It is sometimes mixed with plant material and feathers (black nests) or forms the entire nest (white nests). These nests are collected by men who climb rickety bamboo scaffolding or vine ladders to reach the high ceilings of caves where tens of thousands of these swiftlets nest. Although white nests are considered the most valuable, as the main ingredient of bird's nest soup, both white and black nests are harvested and have become a major economic resource in that part of the world. Harvesting is controlled to protect the birds and keep this a renewable resource. Occasionally swiftlets, as well as other species of swifts, nest in close association with humans and use buildings and bridges as nest sites; chimney swifts now nest more commonly in chimneys than in hollow trees.

Researchers have found that several species of swifts and their nestlings are able to survive short periods of inclement weather by entering a semi-torpid state with a lowered body temperature. There are anecdotal accounts of what appears to be true hibernation in the chimney swift, but this needs further study.

Some cave swiftlets (genus *Aerodramus*) can nest and roost in total darkness deep in caves, sometimes more than a kilometer from any light. These birds make a series of audible clicks or rattle calls, and the returning echoes enable them to navigate within a cave and locate their own nest or roost site. (The only other bird that uses echolocation is the oilbird *Steatornis caripensis* of northern South America and Trinidad.) Non-echolocating swiftlets nest in the twilight zone and near the entrance of caves where there is still sufficient light for visual flight.

Both sexes participate in nest building,

incubation, and provisioning of the chicks; incubation requires 19 to 23 days, and the nestling period may last as long as six to eight weeks. For species that nest in colonies there often is much social activity in the form of grouped flights and vocalizing. Their calls vary, from short sharp chips to long-drawn-out buzzy screes or screams.

CRESTED SWIFTS

The four species of crested swifts (genus *Hemiprocne*) are distributed from peninsular India eastward through Malaysia and the Philippines. All have frontal feathered crests, various degrees of forked tails, and patches of brighter colors. They are far less aerial in their behavior than the true swifts, in that often they alternate between perching on prominent treetops and making graceful flights in pursuit of flying insects. Nowhere as abundant as swifts, crested swifts tend to be more solitary and sparsely distributed. Their nest is tiny, consisting of a small cup of plant material and lichens barely large enough to hold the single egg; it is fastened to a small lateral twig, and the brooding bird straddles the nest and supports itself on the underlying branch. Because their nests are typically high in the outer branches of large trees, many aspects of the breeding biology of crested swifts remain to be studied.

HUMMINGBIRDS

Hummingbirds are known for their small size, bright iridescent colors, and hovering flight. This diverse New World family, with 320 species in 112 genera, is most abundant in the warm tropical areas of Central and South America, but some are also found from Alaska to Tierra del Fuego and from lowland rainforest to high plateaus in the Andes. The average weight of these tiny birds is between 3.5 and 9 grams (less than ⅓ ounce)—the bee hummingbird is perhaps the smallest living species of bird, at about 2.5 grams (¹⁄₁₀ ounce)—but a few are larger, the giant hummingbird *Patagona gigas* being almost 20 grams (⅔ ounce).

The shape of the bill clearly reflects the type of flowers each species visits for nectar and insects. Hummingbird foraging takes two major forms: territoriality, in which floral nectar sources are vigorously defended; and trap-line foraging, where rich but more widely dispersed sources of nectar are regularly visited. The tongues of hummingbirds are brush-tipped to aid in nectar acquisition, but insects provide a needed source of protein and are a major component of their diet.

The extremely rapid wing beat (22 to 78 beats per second), coupled with a rotation of the outer hand portion of the wing and a powered upstroke, permits hummingbirds to hover adroitly in front of flowers during foraging. They also make vigorous acrobatic flights during territorial chases, and some species make elaborate aerial courtship displays. Longer flights to follow seasonal flowering patterns

▼ *A male and a female black-chinned hummingbird consort briefly at the same flower. In most hummingbirds there is no pair bond, and the sexes come together only casually or when actively courting.*

Bob & Clara Calhoun/Bruce Coleman Ltd

Robert A. Tyrrell

are also undertaken, and some species migrate over several thousand kilometers from nesting areas in temperate zones to wintering grounds in the tropics. During its migration, the ruby-throated hummingbird *Archilochus colubris* flies 1,000 kilometers (620 miles) across the Gulf of Mexico.

The breeding season of most species is keyed to the local flowering cycle, although it avoids seasons of intense rain. In most species the female alone builds the nest and incubates the eggs. Nests are typically small cups of plant material held together with spider web and sometimes adorned with moss or lichen. Bulkier nests, sometimes attached to the underside of a leaf, are typical of the hermit hummingbirds and some cave-nesting species.

CHARLES T. COLLINS

▲ *A hummingbird drains the nectar from a flower. Hummingbirds need constant access to flowers to fuel their prodigious metabolism, by far the highest measured among birds. An intricate relationship exists between many hummingbirds and the flowers at which they feed, and some flowers rely on hummingbirds for pollination.*

MOUSEBIRDS & TROGONS

Mousebirds (order Coliiformes), also known as colies, are drably colored, small-bodied birds of Africa. The name "mousebird" comes from the curious way in which the birds creep and crawl among the bushes, clinging upside down, with the long tail high in the air. Among the trogons (order Trogoniformes) the resplendent quetzal *Pharomachrus mocinno* of Central America is perhaps the best known, but all of these birds of the tropical woodlands and forests have brilliantly colored plumage.

MOUSEBIRDS, OR COLIES

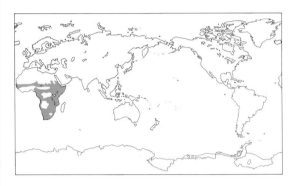

Mousebirds are distributed throughout Africa south of the Sahara. The species inhabit a wide range of country, from almost-dry bushland to the edge of forests (but not in the forest), from low altitudes to 2,000 meters (6,500 feet). Some have taken up residence in people's gardens and smallholdings, where they are justifiably regarded as pests. Mousebirds are vegetarians and their diet consists of foliage, buds, flowers, wild or cultivated fruits, and even seedlings.

The nest is well hidden in a bush, and is of rough construction but has a softly lined cup. The

► Confined to Africa, the mousebirds are superficially unremarkable in appearance, but they have a number of odd characteristics. Like the other species, the speckled mousebird habitually perches in a distinctive fashion with the feet more or less level with the shoulders. Its feathers have very long aftershafts, which contributes to an unusually fluffy-looking plumage.

John Shaw/NHPA

eggs (usually four to six) are laid at daily intervals, and incubation begins with the first, with the result that nestlings of different sizes will be found. The nestlings are naked when newly hatched but soon grow down and then feathers. Older nestlings, if disturbed or in full sun, may clamber out and hide—but then may be cannibalized by their parents. Nest building, incubation, and feeding by regurgitation are undertaken by both parents. A family may remain together for a considerable time or may join others to form groups, the leader going from bush to bush while others follow. Their flight is often swift and direct, with whistle contact-calls. When resting or roosting the family or group will clump together haphazardly, some clinging to vegetation, others on the backs of those below. Their legs and feet are curiously articulated.

Mousebirds dislike rain and cold, and even huddling together they may become torpid. They seldom drink, but groups are often seen dust-bathing and sunning themselves with wings and tail well spread. The feathers grow randomly, and the birds are often parasitized by flies, lice, fleas, and ticks.

TROGONS

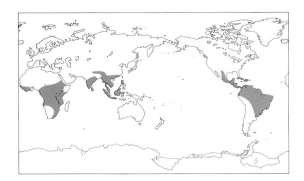

Trogons are pantropical birds of Central America and the West Indies, Africa, and Asia. They are forest dwellers, not very well known, and there is still disagreement among taxonomists about details of their classification. Generally the head, breast, and back are metallic iridescent green with some reflections of blue and yellow. The belly may be carmine, red or pink, orange or yellow. Females are usually duller than the males, and Asian species are not as colorful. The tail is long and slightly graduated, usually with white or black bars. Quetzals (genus *Pharomachrus*) of Central America are well known for their long, drooping upper tail-coverts. The lower part of the leg is feathered, and the toes are heterodactylous (toes 1 and 2 directed backwards, toes 3 and 4 forwards).

Secretive and territorial, a trogon will perch quietly scanning for a food item, especially insects. With a slow undulating flight it may snatch a caterpillar from a leaf or twig, then return to its perch. Occasionally small lizards are taken; and some South American species also eat fruit. The male calls to attract a mate to a suitable nest site.

▶▼ *The national bird of Guatemala, the resplendent quetzal (right) was considered divine by the Aztecs, and killing one was a capital offence. The bird lost this protection after Cortes laid waste the Aztec Empire, and its numbers have declined steadily ever since. It inhabits cloudforest above 1,300 meters (about 4,000 feet); females differ from males most conspicuously in lacking the tail streamers. The violaceous trogon (below) inhabits rainforest edges and clearings from Mexico to Amazonia.*

Females may answer; some remain silent. At each note the tail is depressed. They nest in a cavity in a tree; some excavate nest holes in dead trees. In the studied species, two or three rounded, some-what glossy eggs are laid. Nestlings are naked and helpless when hatched, but soon acquire down. Both parents take part in the incubation and care of nestlings and fledglings. Males may indulge in a display in which several gather to chase each other through the trees. Their skin is delicate and tears readily; soft feathers may fall out.

G. R. CUNNINGHAM-VAN SOMEREN;
REVISED BY JOSEPH FORSHAW

KINGFISHERS & THEIR ALLIES

ORDER CORACIIFORMES
• 9 families • 47 genera
• 206 species

SMALLEST & LARGEST

Puerto Rican tody *Todus mexicanus* and allied species
Total length: 10 cm (4 in)
Weight: 5–6 g (⅙ oz)

Southern ground hornbill *Bucorvus cafer* & Abyssinian ground hornbill *B. abyssinicus*
Total length: 80 cm (31 in)
Weight: 3–4 kg (6½–8¾ lb)

CONSERVATION WATCH
!!! There are 3 critically endangered species: writhed-billed hornbill *Aceros waldeni;* Sulu hornbill *Anthracoceros montani;* Visayan hornbill *Penelopides panini.*
!! 5 species are listed as endangered: silvery kingfisher *Alcedo argentata;* Marquesan kingfisher *Todiramphus godeffroyi;* rufous-lored kingfisher *Todirhamphus winchelli;* writhed hornbill *Aceros leucocephalus;* Mindoro hornbill *Penelopides mindorensis.*
! 17 species are listed as vulnerable.

Kingfishers, todies, motmots, bee-eaters, rollers, hornbills, the hoopoe, and wood-hoopoes are linked by details of their anatomy and behavior. They have generally small feet with a fusing of the three forward toes, and an affinity of the middle ear bone and of egg proteins. Many species are brilliantly colored birds which make their cavity nests by digging holes in earth-banks or rotten trees.

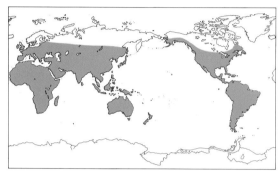

KINGFISHERS

The most visible distinguishing features of the kingfishers (family Alcedinidae) are their bill and feet. The legs are very short, and the toes are syndactyl, the third and fourth being united along most of their length, and the second and third joined basally; in some species the second toe is much reduced or absent altogether. A "typical" kingfisher bill is proportionately large, robust, generally long and straight, with a sharply-pointed or slightly hooked tip.

Two predominant bill shapes reflect the two main ecological groupings of kingfishers: in the subfamilies Cerylinae and Alcedininae the laterally compressed, pointed bill is for striking at and grasping prey, especially fishes, whereas in the subfamily Daceloninae the rather broad, slightly hook-tipped bill is more suited to holding and crushing prey. Some species have intermediate bills, and the bills of a few others are modified for specialized foraging techniques. As its name implies, the hook-billed kingfisher *Melidora macrorrhina* from New Guinea has a broad bill with a strongly hooked tip, apparently an adaptation for digging in the ground to procure prey, whereas the shovel-billed kingfisher *Clytoceyx rex,* also from New Guinea, uses its short, wide, and extremely robust bill to shovel down into moist soil in search of earthworms.

Kingfishers vary in size, from the laughing kookaburra *Dacelo novaeguineae* of Australia or the giant kingfisher *Ceryle maxima* of Africa, with a total length of more than 40 centimeters (16 inches) and weight of about 400 grams

(14 ounces), to the black-fronted pygmy kingfisher *Corythornis lecontei* of Central Africa at 10 centimeters (4 inches) and little more than 10 grams (⅓ ounce). Their plumage is commonly of bright colors, often with a metallic brilliance, in blues, greens, purple and reddish or brown tones which are frequently offset with white patches or dark markings. The wings tend to be short and rounded, while the tail varies from extremely short, as in the *Ceyx* species, to very long in the beautiful *Tanysiptera* kingfishers.

Kingfishers are distributed worldwide, except the highest latitudes and remote islands. The center of abundance is Southeast Asia and New Guinea, but tropical Africa also has strong representation. The northernmost species, summer-breeding migrants, are the Eurasian kingfisher *Alcedo atthis* in southern Scandinavia and western Russia and the belted kingfisher *Ceryle alcyon* in Alaska and Central Canada. The southernmost limit is Tierra del Fuego, which is the southern extremity of the summer breeding range of the ringed kingfisher *C. torquata*. The Eurasian kingfisher is the most widely distributed species with a range extending from western Ireland to the Solomon Islands. Most of the species with very restricted ranges are confined to small islands or groups of islands.

◀ *Kingfishers of the genus* Tanysiptera *are closely related to the widespread* Todiramphus *kingfishers but are characterized by greatly elongated central tail feathers. All eight species inhabit the New Guinea region, but one, the white-tailed or buff-breasted paradise-kingfisher, shown here, migrates across Torres Strait to breed in the extreme northeastern part of Australia. It nests in termite mounds, often close to the ground.*

Tom & Pam Gardner

▲▼ A trio of Old World kingfishers: the white-throated kingfisher of southern Asia from Iraq to the Philippines (left); the spangled kookaburra of New Guinea (right); and the malachite kingfisher of Africa and Madagascar (center).

▼ The Cuban tody is common in forest and woodland, especially along rivers and streams. All five tody species are confined to the West Indies.

D. & M. Zimmerman/Vireo

Kingfishers inhabit a wide variety of aquatic and wooded habitats, though tending to avoid open country where there is a dearth of trees. There are noticeable habitat preferences among fish-eating kingfishers, with some species showing marked preferences for littoral habitats, especially mangroves, whereas others favor large open waterways. Lowland tropical rainforest is the habitat favored by many "non-fishing" kingfishers.

Regular seasonal migration is undertaken by some species occurring at higher latitudes, particularly in the Northern Hemisphere. The belted kingfisher of North America, the Eurasian kingfisher of Europe and western Russia, and the sacred kingfisher *Todiramphus sanctus* of Australia are migrants whose seasonal departures and arrivals are well known. Some continental populations of the Eurasian kingfisher are known to migrate over 1,500 kilometers (930 miles) between breeding and wintering areas.

Because the best-known species, the Eurasian kingfisher, feeds mainly on fish captured by diving into the pond or stream from an overhanging branch, the general concept is that all other kingfishers do likewise. In fact, most species are generalized predators that take a wide variety of invertebrates and small vertebrates from the ground or in water. The usual "sit and wait" foraging technique involves quiet surveillance from a vantage perch, then a swoop or dive down to seize the prey, which is brought back and, while firmly grasped in the bill, is immobilized by being struck repeatedly against a branch. Undigested food is regurgitated as pellets.

Territorial and courtship displays include loud calling and conspicuous flights high above the treetops. The normal breeding unit is a monogamous pair, which may nest solitarily or in loose colonies, but cooperative breeding, involving helpers at the nest, also occurs. With the blue-winged kookaburra *Dacelo leachii* and the laughing kookaburra in Australia, members of the family group participate in all facets of breeding, and more than one female may lay in the same nest.

Nests are in natural hollows in trees or in burrows excavated by the birds in earth-banks, termite mounds, or rotten tree-stumps. Both sexes excavate the burrow by prizing loose the soil or rotten wood with the bill, and then by using their feet to kick out the dislodged material. At the end of the burrow there is an enlarged chamber, where the eggs are laid on the unlined floor. During the course of nesting, the burrow becomes fouled with excrement, regurgitated pellets, and food remains, resulting in a characteristically pungent odor. Chicks of small fish-eating species such as the Eurasian kingfisher become independent almost immediately after fledging, whereas chicks of the larger woodland kingfishers are dependent on their parents for up to eight weeks after leaving the nest.

TODIES

Todies are very small, stocky birds with a proportionately broad head, fairly long, much-flattened bill and a short, slightly rounded tail. Their striking plumage has brilliant green upperparts and an almost luminescent red throat-patch. The sexes are alike, but young birds are noticeably duller than adults and lack the red throat-patch or have it only slightly indicated. Differences in the color of the flanks and the presence or absence of bluish subauricular patches are the main distinguishing features of the species.

All five species belong to the genus *Todus* and are confined to the West Indies, where the only island with two species is Hispaniola: the narrow-billed tody *T. angustirostris* in wet montane forests, and the broad-billed tody *T. subulatus* in dry forests or scrublands at a lower altitude. On other islands the single species shows a wider habitat tolerance. Todies remain common throughout much of their restricted range and seem to have coped remarkably well with the two major threats that affect so many island birds, loss of habitat and introduction of predators or competitors.

The typical method of capturing their insect prey is to take it from the underside of a leaf in the course of an arc-like flight from one perch to

D. Wechsler/Vireo

another. The flattened bill seems well suited to this technique and also to catching insects in flight. The nest is in a burrow excavated by the birds in an earth bank, and the two or three white eggs are laid in a slight depression in the floor of an enlarged chamber at the end of the burrow. Incubation is shared by the sexes, and chicks are fed by both parents. At some nests, extra adults have been recorded assisting in the feeding of nestlings.

MOTMOTS

The 10 species in the Momotidae family are confined to continental tropical America, from northern Mexico to northern Argentina. They are birds of forests and woodlands in the lowlands and foothills, a notable exception being the blue-

▲ Despite its colorful plumage, the rufous motmot is an unobtrusive bird of dense rainforest, more often heard than seen. It perches quietly, capturing prey on the ground or near it in brief sallies. Most prey is brought back to the perch to be subdued and swallowed. Motmots have a distinctive habit of persistently swinging the tail from side to side like a pendulum.

Hilary Fry

▲ In typical bee-eater stance, a cinnamon-chested bee-eater sits alert for prey. This species is common in the highlands of East Africa.

throated motmot *Aspatha gularis* which is restricted to cloud forest and mixed woodlands in the highlands of Central America.

Motmots are robust birds with a distinctive overall appearance that comes from a broad head with a proportionately long, strong bill, as well as proportionately long legs and, in most species, an acutely gradated tail with elongated central feathers. Variation in size is quite marked, from the tody motmot *Hylomanes momotula*, total length 19 centimeters (7½ inches) to the upland motmot *Momotus aequatorialis*, total length 53 centimeters (21 inches). The plumage is a combination of green and rufous, and most species have striking head and facial patterns in black and shades of brilliant blue. The stout, slightly decurved bill varies from rounded and laterally compressed to extremely broad and flattened, and in all species except the tody motmot the cutting edges are serrated.

It is the tail that most typifies the motmots, for in some species the elongated central tail-feathers are denuded part-way along their length but end in fully-intact tips that form flaglike spatules. Newly-acquired feathers are without these bare shafts, and it may be that the barbs are loosely attached and therefore fall away as the bird preens and also in consequence of rubbing against vegetation. The tails of females are slightly shorter than those of males, but otherwise the sexes are similar. Young birds resemble adults, though the head and facial markings are usually duller.

Motmots fly forth from vantage perches to capture prey on the ground, amid foliage, from the surface of a branch or tree trunk, or in the air.

Usually, captured prey is brought back to be struck repeatedly against the perch before being swallowed. Motmots have been seen with other birds following army ants to capture insects fleeing from the marauding columns.

The nests are burrows excavated by the birds in earth-banks, and some species nest in colonies of just a few pairs or 40 or more pairs. The eggs are laid in an enlarged chamber at the end of the burrow, and the incubation (about 20 days) is shared by the sexes. Both parents feed the chicks, which leave the nest about 30 days after hatching. Motmots remain locally common in much of their range, but their future is uncertain as the destruction of forests and woodlands continues unabated.

BEE-EATERS

Bee-eaters are particularly graceful birds, both in appearance and in their actions. A slim body shape is accentuated by a narrow, pointed bill and a proportionately long tail, the central feathers of which are elongated in many species. Bright green upperparts are a prominent feature of most species, and there are distinctive head or facial patterns. The predominantly red plumage, with little or no green, of the northern carmine bee-eater *Merops nubicus*, southern carmine bee-eater *M. nubicoides*, and rosy bee-eater *M. malimbicus* clearly sets apart these three spectacular species. In all species the sexes are alike or differ very slightly, and young birds are generally duller than adults. In size the adults vary from 18 to 32 centimeters (7 to 12½ inches).

Being specialist predators of stinging insects,

▶ The swallow-tailed bee-eater is common in scrub and savanna woodland across much of Africa. Markedly less gregarious than some bee-eaters, it is usually encountered in pairs or small parties. Bee-eaters feed on flying insects such as bees and butterflies, capturing them in marvelously dextrous sallies from high exposed perches.

Nigel Dennis/NHPA

◄ The blue-crowned motmot (far left) is the most widespread member of this small American family of birds; in Brazil it is known locally as the hudu, expressive of its most persistent call. The rainbow bee-eater (near left) is the only Australian representative of its family, but it is common and widespread across much of the continent. It is strongly gregarious, and most populations are migratory.

bee-eaters have developed an effective technique for devenoming their prey. A captured insect is struck repeatedly against the perch, and then is rubbed rapidly against the perch while the bird closes its bill tightly to expel the venom and sting. Bees, wasps, ants, and their allies predominate in the diet of most species, and insects nearly always are captured in the air. As specialized insectivores, bee-eaters may be vulnerable to the effects of chemical pesticides, although no species are at present in danger of extinction.

The 27 species occur only in the Old World, where the center of abundance is in northern and tropical Africa. Most live in open, lightly wooded country, often in the vicinity of waterways. Along routes through the Middle East and the Arabian Peninsula, a conspicuous seasonal phenomenon is the passage of large numbers of the European bee-eater *Merops apiaster* and the blue-cheeked bee-eater *M. persicus* migrating between their breeding areas in Europe and Asia and the wintering grounds in Africa. There are also intracontinental migrations in Africa. Bee-eaters are common in Southeast Asia, where there is not only seasonal movement over shorter distances but also long-

distance regular migration—for example, the rainbow bee-eater *M. ornatus* which reaches Australia for breeding in spring.

The burrow nests excavated by bee-eaters in an earth-bank or flat sandy ground may be solitary (as for the *Nyctyornis* species), in groups of two or three (as for the black-headed bee-eater *Merops breweri*), in loose aggregations spaced regularly or irregularly along a bank, or in large colonies containing up to a thousand or more holes. Incubation is shared by the sexes and both parents feed the chicks, which vacate the nest 24 to 30 days after hatching. The role of additional adults as helpers at the nest has been confirmed for some species and is suspected to be present in others—it has been determined that participation by helpers increases the fledging success rate quite dramatically.

ROLLERS
Rollers take their name from the twisting or "rolling" actions that characterize the courtship and territorial flights undertaken by some of the 12 species in the family Coraciidae. Typical rollers are fairly large stocky birds, 25 to 37 centimeters

(10 to 14½ inches) long, with robust bills and rather short legs. The squarish tail is proportionately short, though some African species have tail streamers. Different foraging techniques are reflected in the bill shape: the eight *Coracias* species capture prey largely on the ground, and their projecting bills are narrower and laterally compressed; the four *Eurystomus* species take insects in the air, and their short, flattened bills are very broad at the base. Bills are blackish in *Coracias,* but bright yellow or red in *Eurystomus.* Plumage is brightly colored, with shades of blue predominating, and young birds resemble the adults, though are noticeably duller.

The European roller *Coracias garrulus* is a regular intercontinental migrant, leaving its breeding range in Eurasia to overwinter in Africa. Rollers are especially prevalent in Africa, where both resident and overwintering species are conspicuous in open and lightly wooded habitats. They range east to Southeast Asia, where there are markedly fewer species, and only one species extends to Australasia, the red-billed roller or dollarbird *Eurystomus orientalis;* the Australian populations of this species migrate north to overwinter in New Guinea.

Raucous calling, bright plumage, and frequent aerobatics make rollers very conspicuous. They are particularly active and vocal at the onset of breeding, and often dive at intruders, including humans, in the vicinity of the nest. From a low vantage perch a *Coracias* roller will drop to the ground to capture a large insect or small vertebrate, whereas *Eurystomus* rollers sally forth from high exposed branches to take flying insects.

Nests are in hollows in trees or in burrows excavated by the birds in earth banks, or

▲ One of the most conspicuous of African birds, the lilac-breasted roller is usually encountered in pairs in open country, often perched high in a dead tree.

sometimes in crevices under the eaves of houses or in adobe walls. Both sexes incubate the eggs for 17 to 20 days and feed the chicks in the nest for up to 30 days.

There have been dramatic declines in numbers of the European roller in many parts of Europe,

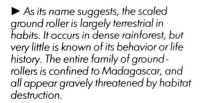

► As its name suggests, the scaled ground roller is largely terrestrial in habits. It occurs in dense rainforest, but very little is known of its behavior or life history. The entire family of ground-rollers is confined to Madagascar, and all appear gravely threatened by habitat destruction.

O. Langrand/Bruce Coleman Ltd

Belinda Wright

possibly because of loss of woodland habitat and the widespread use of chemical pesticides. Two species, the azure roller *Eurystomus azureus* and Temminck's roller *Coracias temminckii,* have very restricted ranges in Southeast Asia and so are vulnerable to habitat disturbance.

GROUND ROLLERS

A group of highly-specialized, terrestrial ground rollers is found only in Madagascar. The five species are stout-bodied birds, varying in length from 25 to 40 centimeters (10 to 15¾ inches), and are immediately distinguished from other rollers by their long sturdy legs, obviously an adaptation for life on the ground. Plumage coloration is variable, but in most species is characterized by finely striped or flecked patterns. Four species frequent the eastern rainforests of Madagascar, where a preference for secluded shaded areas, coupled with their secretive habits, make detection difficult, and so they remain poorly known. The long-tailed ground roller *Uratelornis chimaera* inhabits arid woodland in the southwest and is common in areas of undisturbed habitat. All are insect-eaters, foraging quietly in the leaf litter, and some species are known to feed at dusk and into the night. The birds excavate burrow nests in the ground, but little is known of their breeding behavior.

THE COUROL, OR CUCKOO-ROLLER

The courol *Leptosomus discolor* is confined to Madagascar and the Comoro Islands. It is a fairly large bird, about 50 centimeters (20 inches) long and weighing up to 240 grams (8½ ounces). There is a marked difference in plumage coloration of the sexes: in the adult male the upperparts are grayish-black, strongly glossed with metallic green and mauve-red, while the underparts and prominent frontal crest are ash-gray; the adult female is almost entirely rich rufous, broadly barred and spotted with black. Young birds resemble the adult female.

The courol is generally common in forests and wooded areas, and is usually encountered singly or in pairs during the breeding season but in groups of a dozen at other times of the year. The birds are confiding, generally allowing a close approach, and their presence is particularly noticeable when they call loudly while flying back and forth above the treetops. Amid the branches of trees and shrubs, they capture large insects and small reptiles, especially chameleons. The nest is in a hollow limb or hole in a tree, where up to four white eggs are laid.

THE HOOPOE

A large erectile crest and a long decurved bill are unmistakeable features of the hoopoe *Upupa epops,* the sole member of the family Upupidae. The

▲ *The Indian roller perches on telephone wires or fence posts, periodically descending, parachute-like, to snatch prey on the ground. It has been known to snatch fish from the surface of water, or even dive in after them like a kingfisher. Rollers are named for their flamboyant aerial courtship displays, which often involve both sexes.*

147

plumage is pink-buff to rich cinnamon or rufous, relieved by black and white banded wings and tail. The adult bird weighs about 55 grams (2 ounces) and is about 28 centimeters (11 inches) long.

Widely distributed in Africa, Eurasia, and Southeast Asia, the hoopoe inhabits a variety of lightly timbered habitats where open ground and exposed leaf litter are available for foraging. It feeds mainly by probing with the long bill into soft earth, under debris, or into animal droppings for insects and their larvae—even in grazing pastures and well-watered gardens. During the breeding season it is usually encountered singly or in pairs, but at other times in family parties or loose flocks of up to 10 birds. The hoopoe is migratory in northern parts of its range, and partially migratory or resident elsewhere. Most migrants from Europe probably overwinter in Africa south of the Sahara Desert, where they are not distinguishable in the field from local residents. Populations from central and eastern Siberia migrate to wintering grounds in southern Asia.

Hoopoes nest in hollow limbs or holes in trees, in burrows in the ground, crevices in walls of buildings, or under the eaves of houses, and will use nest-boxes. They reuse some sites year after year. Only the female undertakes incubation, which lasts 15 to 20 days. Newly-hatched chicks are brooded and fed by the female, but later are cared for by both parents. The nesting hollow becomes fouled with excrement and remnants of food as the chicks grow, and eventually is vacated by the brood some 28 days after hatching. Though the species remains generally common throughout much of its range, there has been a steady decline in numbers in Europe during most of this century.

WOOD-HOOPOES

Endemic to Africa south of the Sahara Desert, the eight species of wood-hoopoe (family Phoeniculidae) are small to medium-sized arboreal birds with long tails and proportionately long, pointed bills, which in some species are strongly decurved. The plumage is black, usually with pronounced iridescent sheens of metallic green, blue, or violet, generally with white bars or spots on the wings and tail, and in some species the head is white or brown. Females usually have shorter bills and tails than the males, and young birds are noticeably duller than adults.

The forest wood-hoopoe *Phoeniculus castaneiceps* and the white-headed wood-hoopoe *P. bollei*, which is larger, favor primary and secondary forest, where they keep to the higher branches and canopy. Other species inhabit mainly savanna, open woodland, and dry thornbush country. All are active birds and clamber along tree trunks or amid foliage with much agility, continually probing in crevices or under bark with their long bills in search of insects and their larvae. As with woodpeckers and woodcreepers, the tail is used as a brace, and often a bird will hang upside down to reach its prey. Larger species tend to be gregarious, noisy, and conspicuous, especially when displaying the white markings on their wings and tails as they fly from one tree to the next.

Wood-hoopoes nest in hollow limbs or holes in trees, often holes originally excavated by barbets or woodpeckers. The purple wood-hoopoe *P. purpureus* is a cooperative breeder, with up to 10 helpers being recorded at a nest; usually these are previous offspring from the nesting pair. Incubation lasts 18 days and is only by the female, then the young birds fledge at 28 to 30 days. The nesting habits of other species are largely unknown.

JOSEPH FORSHAW

HORNBILLS

Hornbills are one of the most easily recognized families of birds, the Bucerotidae. They are conspicuous for their long down-curved bills, often with a prominent casque on top, their loud calls and, at close range, their long eyelashes. All but two species are also notable for sealing the entrance to their nest cavity into a narrow vertical slit, the incarcerated female and chicks being fed by the male. Hornbills are confined to the Old World regions of sub-Saharan Africa and of Asia east to the Philippines and New Guinea. Toucans, some with relatively larger and more gaudy bills than hornbills, are their New World equivalents.

The large bill of hornbills serves a variety of functions including feeding, fighting, preening, and nest-sealing. The head is supported by strong neck muscles and strengthened by fusion of the

W.S. Paton/Bruce Coleman Ltd

◄ Widely distributed in Africa, the purple wood-hoopoe is a noisy, gregarious species that lives in small parties. The birds move through forest and woodland, following each other from tree to tree. The wood-hoopoe feeds mainly on insects, using its curved bill to probe for them in crevices in the bark. It climbs actively about limbs and branches, often clinging to the trunk using its tail, woodpecker-like, as a prop.

first two neck vertebrae, unique among birds. The casque, barely developed in species such as the southern ground hornbill *Bucorvus cafer* and red-billed hornbill *Tockus erythrorhynchus,* appears to function primarily as a reinforcing ridge along the crest of the bill. However, in many species it is modified secondarily: as a hollow resonator in the black-casqued hornbill *Ceratogymna atrata;* as a special shape in the wrinkled and the wreathed hornbills *Aceros corrugatus* and *A. undulatus;* or, most extremely, as a block of solid ivory in the

◄ *(Opposite page) A hoopoe arrives at its nest. Widespread in the warmer parts of Europe, Africa and Asia, the hoopoe nests in cavities in trees, among rocks, or burrowed in earthen banks. The female incubates alone while food is brought to her by the male. After the young hatch, the female joins the male in bringing food to the nest, but normally the male merely fetches, handing the food to the female who in turn presents it to the young.*

◄ A yellow-billed hornbill. All but one of the 14 species in the genus Tockus are confined to Africa. They differ from most other hornbills in their ground-foraging habits, and in feeding largely on insects rather than fruit.

helmeted hornbill *Rhinoplax vigil,* so that the skull is 11 percent of the body weight.

The plumage of hornbills consists of areas of black, white, gray, or brown, with very few special developments apart from a loose crest, as in the bushy-crested hornbill *Anorrhinus galeritus,* or long tail feathers as in the helmeted hornbill. However, the bill, bare facial skin, and eyes are often brightly colored, with reds, yellows, blues, and greens, and the throat skin may form wattles, as in the yellow-casqued hornbill *Ceratogymna elata,* an inflated sac, as in the wreathed or ground hornbills of the genus *Bucorvus,* or the whole head and neck may be exposed as in the helmeted hornbill.

In all species, development of the bill and casque indicates the age and sex of individuals, supplemented in many species by differences in colors of the plumage, bare facial skin, or eyes. Males are slightly larger than females, and the shape of their more prominent casques is often indicative of the species, such as in the rhinoceros hornbill *Buceros rhinoceros.* Each species also has loud and distinctive calls—among the most obvious in their environment—from the clucking of von der Decken's hornbills *Tockus deckeni* or the booming of Abyssinian ground hornbills *Bucorvus abyssinicus* in Africa, to the roars of great Indian hornbills *Buceros bicornis* or the hooting and maniacal laughter of the helmeted hornbill in Malaysia. Several species conduct conspicuous displays while calling, and most larger species are audible in flight, the air rushing between the base of their flight feathers where lack of stiff underwing coverts is another feature of the family.

▶ *Largest of all the hornbills, the African ground hornbill lives in open country in small family parties that daily patrol well-defined territories. Maintaining contact with frequent loud, deep booming notes, the birds spread out and methodically search through grassland, feeding on a wide variety of ground-living animals, including large insects, snakes, and frogs.*

Mitch Reardon/Weldon Trannies

Stanley Breeden

The 45 species occupy habitats varying from arid steppe to dense tropical rainforest. All but one of the 12 species that inhabit savanna and steppe occur in Africa, the exception being the Indian gray hornbill *Meniceros birostris*. These include the extremes of size found within the family: from 150 grams (5¼ ounces) to 4 kilograms (8¾ pounds). Most hornbills are sedentary and live as mated pairs within defended territories, which range in size from 10 hectares (25 acres) up to 100 square kilometers (39 square miles). However, species of arid habitats are often forced to move locally during dry seasons, sometimes on a regular migration. Rainforest species may also range widely in search of fruit-bearing trees, especially members of the genus *Aceros* such as the wreathed hornbill, which will daily cross open sea between islands. A number of species reside in groups where all members defend the group territory, and most help a dominant pair to breed.

Hornbills nest in natural cavities in trees or rock faces, and in most species the female closes the entrance hole to a narrow vertical slit, using the side of the bill to apply mud, droppings, or food remains as a sealant. In some species, such as the great Indian hornbill, the male assists with delivering mud and sealing the exterior; and in others the male swallows mud to form pellets in his gullet which are then regurgitated to the female. The entrance is also resealed by half-grown chicks in those species where the female emerges before the end of the nesting cycle. The exceptions are the two species of ground hornbill, which neither seal the nest nor squirt their droppings out of the entrance for sanitation, and which may excavate their own holes in earth-banks or even use old stick nests of other birds.

In all species the male delivers food to his mate, and later to the chicks, assisted by group members in cooperative species. The food is carried as single items in the bill tip of small insectivorous species, as a bolus of several small animals in the carnivorous ground hornbills, or as a load of fruits in the gullet of many forest species, which are then regurgitated one at a time at the nest and passed in to the inmates.

Small hornbills lay up to six eggs, have an incubation period of about 25 days, see the female emerge when the eldest chick is about 25 days old, and have a nestling period totaling about 45 days. Large species lay two eggs, incubate for about 45 days, and leave the chick alone when it is a month old even though the nestling period extends to about 80 days. In most large forest species the female remains in the nest until the chick is fledged, a total period of incarceration of four or five months. Females of most species molt while in the nest, casting all flight feathers within a few days of beginning to lay eggs, and regrowing them before emerging from the nest.

ALAN KEMP

▲ *The great Indian hornbill feeds largely on fruit, especially figs; these are usually plucked from among the foliage, but the bird will sometimes descend to the ground to gather fallen fruit. Most often found in small parties that roost communally, this hornbill habitually visits several fruiting trees in succession on a regular daily basis.*

ORDER PICIFORMES
• 6 families • 65 genera
• 378 species

SMALLEST & LARGEST

Asian rufous piculet *Sasia abnormis* & South American scaled piculet *Picumnus squamulatus*
Total length: 77 mm (3 in)
Weight: 7 g (¼ oz)

Black-mandibled toucan *Ramphastos ambiguus* & white-throated toucan *R. tucanus*
Total length: 61 cm (2 ft)
Weight: 500 g (17½ oz)

CONSERVATION WATCH
!!! The imperial woodpecker *Campephilus imperialis* and Okinawa woodpecker *Sapheopipo noguchii* are critically endangered.
!! 4 species are listed as endangered: three-toed jacamar *Jacamaralcyon tridactyla*; white-mantled barbet *Capito hypoleucus*; Cuban flicker *Colaptes fernandinae*; helmeted woodpecker *Dryocopus galeatus*.
! There are 8 vulnerable species.

WOODPECKERS & BARBETS

The order Piciformes contains six families of cavity-nesting birds: jacamars (family Galbulidae), puffbirds (Bucconidae), barbets (Capitonidae), toucans (Ramphastidae), honeyguides (Indicoridae), and woodpeckers (Picidae). Although they look different, they share certain anatomical features such as zygodactyl feet (two toes in front, two behind) with associated tendons that flex them, they generally lack down feathers (except Galbulidae), and lay white eggs. Most species live in tropical regions. Most are colorful, and some are very gaudy.

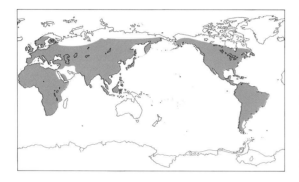

BILLS FOR SPECIAL PURPOSES
The largest family (with more than half the 378 species) are the well-known woodpeckers, which are distributed around the world except on oceanic islands, Australasia, Madagascar, Antarctica, and Greenland. Woodpeckers and barbets excavate roosting and nesting cavities, which are later taken over and used by hole-nesting species of other families and orders. Hence, many other birds depend on them.

There is much variation in feeding habits and therefore in bill structure. Jacamars have longish, pointed bills and are insect-eaters. Puffbirds too are insectivores, but have large, often hooked bills. Barbets generally are fruit-eaters, with a largish, sometimes notched bill. Toucans have remarkably large colorful bills, somewhat resembling those of hornbills, and feed mainly on fruits. The woodpeckers are unique in their strong, tapering, often chisel-tipped bills and strong tail feathers used as a prop against the bark of trees, where they glean, pry, probe, and excavate for insects. Honeyguides eat insects and particularly wax obtained from beehives

► A purple-necked jacamar. Jacamars are exclusively American, while the bee-eaters are restricted to the Old World; although the two groups are unrelated they show some remarkable similarities in diet, appearance, and general behavior.

J. Dunning/Vireo

◄ The gaudy red-and-yellow barbet (far left) lives in groups in East African scrublands, nesting in termite mounds. The rufous-tailed jacamar (near left) is one of the most widespread members of its family, inhabiting clearings, farmland, and forest edges from Mexico to northern Argentina.

or the exudate of certain insects, using their relatively unspecialized short bills; one African species "guides" humans to the location of beehives, giving rise to the name of the family.

JACAMARS

The male rufous-tailed jacamar *Galbula ruficauda* (length 25 centimeters, or 10 inches) is long-billed and long-tailed, metallic green in color with elongated central tail feathers, and generally rufous or chestnut below with a white throat. The female is similar but buff-colored below and on the throat. They live a solitary life except when breeding. Then they excavate a nesting tunnel in a termite mound or earthen bank, and for about three weeks both parents incubate the eggs (two to four) during the day, the female incubating alone at night. The young hatch bearing whitish down feathers, and are fed insects, especially butterflies, which the parents catch by flying out from a perch. Nestlings make trilling calls, weak versions of the adults' notes. When leaving the nesting cavity about 24 days after hatching, the fledglings closely resemble their parents, sex for sex. This species is less forest-dependent than most other members of this strictly American family, and it may be found even in grassland with scattered trees. Perhaps that explains its very extensive distribution from Mexico to Argentina.

Some ornithologists consider that the jacamars

and the puffbirds are not related directly to the other four families in the order Piciformes.

PUFFBIRDS

Like jacamars, the puffbirds sally forth from a perch to catch insects, but in contrast to the giant hummingbird-like appearance of the colorful jacamars, puffbirds are stout-billed, short-necked, big-headed birds of subdued colors. Most are solitary species, but the nunbirds (genus *Monasa*) are social and even breed cooperatively. The black nunbird *M. atra* digs a tunnel nest into the ground, then covers the entrance with a pile of sticks, under which a horizontal "tunnel" leads to the actual nest chamber; adults in a cooperatively breeding group utter a loud chorus of "churry-churrah" notes, answered by neighboring groups. The swallow-winged puffbird *Chelidoptera tenebrosa,* a small-billed, mainly black bird (white rump, rufous belly) of northern South America, makes long flycatching sweeps, often repeated, hence is quite aerial. Its nest is excavated straight into the forest floor and lacks a stick-covering or "collar" of twigs or leaves around the entrance.

BARBETS

Tropical Asian, African, and American forests and woodlands and (in Africa) arid scrublands, are home to the brightly colored and patterned barbets. These

Belinda Wright

▲ Not much bigger than a house sparrow, the coppersmith barbet is a colorful, arboreal bird that obtains most of its insect food from the bark of trees. It takes its name from its monotonous, metallic call-notes.

of the Andes, from Colombia to Bolivia. Females are duller, with a whitish crown. Repetitive croaking notes are uttered when breeding, but generally they are inconspicuous.

African barbets are the most diverse, ranging from highly social (up to 50 or more pairs nesting in one dead tree), dull-colored brown or olive species, to the group-social, duetting, multicolored species of the genus *Lybius,* and the solitary, tiny (length 7 to 8 centimeters, or 3 inches) tinkerbirds of the genus *Pogoniulus.* Unusual among them are the ground-barbets which nest in holes excavated straight into the ground, or in termite mounds. The gaudy red-and-yellow barbet *Trachyphonus erythrocephalus* (length 20 centimeters, or 7¾ inches) is marked with black, yellow, white and red spots, stripes and blotches. Nesting in groups, they excavate holes in banks or termite mounds. A group of five to eight chorus together regularly, all year, uttering a melodious series of notes sounding like a repetitive "red-'n-yell-oh", but the primary pair maintain the highest perches among the group, carry on displays with the head and tail, and sustain the chorus as a duet of their own. The bright orange-red bill is long and tapering, and with it this barbet seizes insects, fruits, and even young birds it encounters in the scrublands of Ethiopia south to Tanzania.

Tropical Asian barbets tend to be larger than those in Africa or tropical America and are large-billed,

have generally heavy bills, some with a notch on either side, with which they seize and bite into fruits such as figs. They excavate holes in dead trees with what at first glance appears to be an ineffective bill that is broad but pointed. A red cap, white facial stripes, olive back, and orange throat mark the male scarlet-crowned barbet *Capito aurovirens* (length 19 centimeters, or 7½ inches), which lives along the base

L.C. Marigo/Bruce Coleman Ltd

▶ A rufous-necked puffbird. Named for their unusually loose, fluffy plumage, the puffbirds are rather inconspicuous, inactive, solitary birds of the middle levels of South American rainforests.

large-headed, short-tailed birds. The numerous species generally are green with sexually-differing intricate patterns of reds, blues, yellows, and blacks concentrated on the head and throat. Most belong to the genus *Megalaima* and are solitary (though they may join other barbets, often of several species, to feed in a fruiting tree), but highly vocal, uttering repetitive ringing "towp", "chook", or "pop" notes during much of the year. The coppersmith barbet *M. haemacephala* is named for its ringing, oft-repeated notes—it is one of several diverse species called "brain-fever birds" because of their monotonous unceasing song. Pairs excavate a nesting cavity in a dead tree trunk or branch, often on the sloping underside, in which they lay their eggs and raise their young.

TOUCANS

The huge-billed, comical-looking toucans of tropical America might be termed "glorified barbets", but of course their large size, frilled tongue, colorfully patterned and often serrated bill, and their inability to excavate a nest-cavity on their own, render them distinctive. Most are black, blue, green, brown, yellow or red on the body, with brilliantly colored stripes or patches on the bill, and often bright, bare skin around the eyes. They adroitly feed on fruits, moving long distances from feeding tree to feeding tree, but can also pluck untended eggs or baby birds out of nests. In flight they appear ungainly and conspicuous, but like parrots they are surprisingly inconspicuous when perched, unless they are calling or displaying.

The collared aracari *Pteroglossus torquatus* (length 41 centimeters, or 16 inches) lives in pairs and social groups, from Mexico to northern South America. Its serrated yellow and black bill is a quarter of its length; the body is greenish above, with a black head, a rufous band around the nape, and bare red skin around the eye, and yellowish below with two red bands, blackish in their center. Fast-flying compared with larger toucans, these aracaris range through forests feeding largely on fruits, calling "ku-sik" repetitively when moving about. They roost in old woodpecker holes, shifting from one to another at intervals, and also nest in such cavities. Sleeping in a group in one "dormitory", they are able to fold the tail over and onto the back. Their blind and naked young find themselves on a pebbly floor of egested seeds after hatching, and both parents as well as up to three helpers join them in the nest for the night.

HONEYGUIDES

Predominantly African in distribution, honeyguides are dull olive-green or grayish birds. Most species have a stubby or pointed bill with raised nostrils, a very thick skin, and an ability to locate beehives. However, only one species, the greater honeyguide *Indicator indicator* (length 20 centimeters, or 7¾ inches), exhibits guiding behavior with a series of calls and flight displays that entice the local people to

◀▼ *Widespread in South American forests, the sulphur-breasted toucan (left) lives in small parties in the treetops. The northern flicker (below) is one of the commonest of North American woodpeckers. It spends much of its time on the ground, feeding mainly on ants.*

follow them to a honey source. The birds are able to locate and enter some hives on their own to feast on beeswax. The male, black-throated and golden shouldered, sings a repetitive song, fending off other males and attracting various pale-throated females with which it mates. The female can breed and lay several clutches a year—using the nests of about 60 diverse host species that have made cavity nests or covered nests—one egg to a nest. She may destroy or remove an egg of the host when she lays; and when the honeyguide egg hatches, the hatchling has a pronounced bill hook with which it regularly strikes out, injuring and killing the hosts' young. When it leaves the hosts' nest the aggressive young bird, in its distinctive yellow and brown plumage, seeks out wax in cavities of trees (abandoned beehives) and soon develops the ability to "guide". Other species of honeyguides monitor noisy humans and their fires in woodlands, watching for the opening of beehives, but they do not "guide" like the greater honeyguide.

WOODPECKERS

Specialists at clinging to the bark of trees and foraging beneath it, woodpeckers excavate several roosting cavities, and a nesting cavity, sometimes into live wood. Their stiffened tail, chisel-tipped bill, bony-and muscle-cushioned skull, and long-clawed strong toes enable them to live anywhere there are trees in

the Americas, Eurasia, and Africa. The smaller species, such as the downy woodpecker *Picoides pubescens* of North America, which is mainly black and white (length 14.5 centimeters, or 5¾ inches), and the great spotted woodpecker *P. major* (length 21 centimeters, or 8¼ inches) of Eurasia, are common visitors to birdfeeding-stations, to the delight of many people who put out suet for them in the cold winter months.

The soft-tailed wrynecks (genus *Jynx*) of Africa and Eurasia are the only members of the woodpecker family that do not excavate their nests; they use natural cavities or old woodpecker or barbet holes. The tiny piculets (genera *Picumnus, Sasia, Nesoctites*) of tropical America, Africa and Asia have short soft tails not used as a brace in "woodpecking", but these woodpeckers drum with the bill to communicate and do excavate their own nesting cavities. Other unusual woodpeckers include some of the flickers (genus *Colaptes*) which live in open country, walk instead of hop, and nest in holes excavated in the ground or in termite nests. Some species of *Melanerpes,* mainly the American acorn woodpecker *M. formicivorus,* are social and nest cooperatively; this species also stores acorns in tiny holes in special storage trees, which are fiercely defended so that the food will be there for the lean times of winter. The sapsuckers (genus *Sphyrapicus*) of North America make rows of small holes in the trunks of certain trees where they regularly feed on the sap that accumulates and on insects attracted to it. Some sapsuckers and flickers are the only members of the order Piciformes that undertake extensive migrations.

Large woodpeckers include the pileated woodpecker *Dryocopus pileatus* (length 42 centimeters, or 16½ inches), widespread in wooded North America. It was decreasing in numbers, but now seems to be adapting and will even nest in trees within sight of New York's skyscrapers. The body is black with white marks; the male has a red crown, crest, and "mustache", whereas the shorter-billed female shows red only on the crest. Members of a pair usually feed alone in a large territory, excavating in rotting trees and fallen logs for tunnels of their chief food, ants, which are extracted by the elongated, frilled-edged, and very sticky tongue. The male and female maintain contact with loud series of "wuk" calls, and they call and drum in a rapid cadence, loudly, to establish and maintain their territory, which is defended sex by sex (the female driving away trespassing females, and the male evicting other males) throughout the year. Each uses several roosting cavities alternately. Both excavate a new nest annually, usually in a partly rotten tree at 10 meters (33 feet) or more above the ground. The male incubates at night, and both alternate by day, during the 18 days to hatching. Adults take turns feeding the young almost hourly, by regurgitating ants, their larvae, and eggs. The young leave the nest at about four weeks of age, and are fed and led about by their parents for two months or more,

Donald D. Burgess/Ardea London Ltd

and then leave or are forced from the territory.

Woodpeckers, especially large species, require large trees and some dead ones. Forest clearing has reduced many populations, and some face extinction. This also threatens other bird species, such as large toucans and hornbills, which partly depend on old woodpecker holes for their own nesting.

LESTER L. SHORT

WOODPECKER'S TONGUE

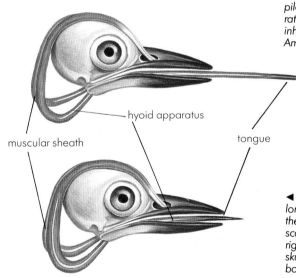

hyoid apparatus

muscular sheath

tongue

▲ Despite its size (about that of a crow or raven) and its vivid red crest, the pileated woodpecker is generally a rather quiet and inconspicuous bird. It inhabits coniferous forests across North America.

◄ The woodpecker's tongue is extremely long and can be extended well beyond the tip of the bill. Sheathed in the scabbard formed by the interior of the right nostril, it loops over and behind the skull to its anchor points in the hyoid bones of the jaw.

◄ (Opposite page) A female great spotted woodpecker leaves her nest in a tree hollow, bearing a fecal pellet. As in many nestling birds, the excrement is ejected wrapped in a gelatinous coating for easy removal by the parents. In woodpeckers these pellets quickly gather a coating of sawdust from contact with the floor of the nest cavity.

BROADBILLS & PITTAS

Among the 5,000 or more species of passerines (or songbirds) grouped in 50 to 74 families—depending upon the opinions of the ornithologists making the grouping—the broadbills are considered the most primitive. They are just over the line dividing passerines from non-passerines. The pittas and two families of primitive species (Philepittidae and Acanthisittidae) are classified vaguely near each other because of the musculature of the syrinx (the organ of voice or song), but geographically and ecologically these groups are widely separated.

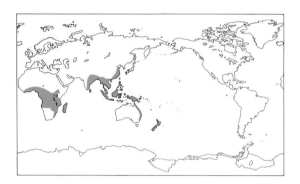

BROADBILLS

The 15 species in the family Eurylaimidae are very colorful, with striking plumage and some having brilliant red, green, or yellow eyes. They are chunky birds with broad heads, broad flattened bills, and short legs. Their feet have three toes forward and one behind. They differ from other passerines in having 15 neck vertebrae instead of 14, and two front toes that may be partially joined at the base; the 11 "typical broadbills" have

▼ *Two colorful birds of Southeast Asian rainforests: the banded pitta (left) searches leaf litter on the forest floor for food, while the green broadbill (right) catches insects in the middle layers of foliage.*

11 primary feathers in each wing, instead of the nine or ten found in other passerines.

The three *Smithornis* species are small dusky birds of forests in Central Africa. During courtship flights they display areas of color; they have loud calls, and hawk insects like flycatchers. The fourth African species is known by only a few specimens from forests near Lake Tanganyika.

All the other species live in India or Southeast Asia. Although most of them are mainly insect-eaters, hawking aerial insects in the lower canopy like clumsy flycatchers or catching them on the ground, they will also take small lizards and frogs. The diet of the three superbly-colored "green broadbills" (genus *Calyptomena*) is predominantly soft fruits and buds. Most species appear to be gregarious and move about the forest in small flocks, which could be family groups. In Malaysia, the long-tailed broadbill *Psarisomus dalhousiae* has been seen in flocks of 20 in association with "bird-waves" totaling more than a hundred birds of seven or eight species. The dusky broadbill *Corydon sumatranus* also joins bird waves; they have been observed sitting quietly on twigs, looking for insects while the flock passes, then flying on to join it and repeating the procedure.

Broadbills' nests are masterpieces of camouflage. The nest is usually attached to a vine suspended in the open, often above a stream, where it appears as debris caught there at high water. This appearance is further amplified by a trailing tail of fibers or debris hanging below it, as well as decorations of lichen and spider webs. Access to the nest hollow is through a hole in the side, which may have a short roof or vestibule. The common Malaysian species, the black-and-red broadbill *Cymbirhynchus macrorhynchos,* which is spectacularly plumaged, nests above streams in dense forest, but also along creeks through farmland. It also suspends its nest above roadsides from power or telephone lines.

ASITYS

The family Philepittidae comprises four poorly known species restricted to Madagascar. They are solitary birds of dense forest undergrowth, small but stout in structure, short tailed, with long legs, and a bill broad at the base and shorter than the head. The velvet asity *Philepitta castanea* of the lowland forests of eastern Madagascar is dimorphic (sexes unalike): the female is greenish-olive, whereas the male has yellow-tipped black body feathers which, when the yellow has worn away, leaves it a velvety black. He also has a long greenish wattle above each eye. The diet appears to be small fruits. They build a bulky nest suspended from twigs in the lower or middle canopy, which has a side entrance with a small projecting roof. Three whitish elongate eggs make up the usual clutch, but incubation and nesting behavior seem not to have been reported.

Schlegel's asity *P. schlegeli,* also a forest species

Morten Strange/Flying Colours

but in western Madagascar, is yellower, and the male is black on the top of the head and has black eye wattles. This species probably has habits similar to those of the velvet asity, but appears to be more active and less restricted to ground cover.

The small-billed false-sunbird *Neodrepanis hypoxantha* is known only by a few specimens from eastern Madagascar. In contrast, the wattled false-sunbird *N. coruscans* is found over much of the island; it is a tiny creeper-like bird only 10 centimeters (4 inches) long, which feeds on insects in bark crevices and also probes flowers with its long downcurved bill. The male is iridescent blue above and yellow below, with a large wattle about each eye. The female is dark green above, yellowish below, and lacks the wattle.

PITTAS

These beautiful birds resemble thrushes—indeed, are sometimes called jewel thrushes—but are separated from them by their syrinx musculature

▲ The black-and-yellow broadbill inhabits lowland forest edges from Burma to Malaysia. In broadbills the sexes are distinguishable but not very different.

159

Glen Threlfo/Auscape International

▲ Despite their brilliant plumage, pittas are surprisingly inconspicuous in the gloom of the rainforest floor. They feed quietly, and seldom call except when breeding. They roost in trees. Here a noisy pitta brings food for its young.

and the structure of the tarsus (leg). The 31 species, all in the genus *Pitta*, are distributed from Africa to the Solomon Islands and from Japan through Southeast Asia to New Guinea and Australia. The probable geographic origin of the species is in the Indo-Malaysian region, which has about 20 species. They are medium-sized insect-eating birds, terrestrial forest inhabitants which, when disturbed, prefer to walk or run rather than fly, yet some species are migratory and travel long distances at night. Most have similar nesting and feeding habits, but the distinctive and vivid coloring of the various species make them some of the world's most brilliant birds. The female is usually duller in color than the male.

Large bulky nests resembling piles of plant debris built on the ground or in low vegetation have a spacious brood chamber and a side entrance

at which the brooding bird rests facing out; the color pattern on the head is such that it blends with the surrounding plant debris, so there appears to be no opening. Both parents care for the eggs and attend the young. All species have loud melodious double whistles, usually heard in early morning and evening.

The African pitta *P. angolensis* breeds in Tanzania and southward to the Transvaal, and when non-breeding migrates north to the Congo basin, Uganda, and Kenya. The green-breasted pitta *P. reichenowi* is a rare forest species of equatorial Africa; it breeds from June into November, and is non-migratory. In Australia, the noisy or buff-breasted pitta *P. versicolor,* a snail-eating forest species, is found from Cape York to New South Wales and breeds from October to January. The northern black-breasted or rainbow pitta *P. iris* favors coastal scrub and mangrove swamps, breeds from January to March, and is thought to be sedentary. The red-breasted pitta *P. erythrogaster* from New Guinea, Java, Malaysia and the Philippines, migrates to northeastern Australia in October; it prefers dense scrub environments and breeds from October to December.

The blue-winged or Moluccan pitta *P. brachyura/moluccensis* (taxonomists are not in agreement as to whether this is one or two species) is the most widely distributed. It is found from India and Sri Lanka east through Southeast Asia to New Guinea and Australia, as well as north through southern China and east into Taiwan, Korea and southwestern Japan. Segments of the population are migratory; pittas that breed in northern India migrate to Sri Lanka, arriving in September and October, overwintering in forests, and leaving in April or May. In Southern Japan they arrive in April or May, breed there, and leave in October. In Malaysia they are permanent residents of mangrove and coastal forests and breed from May into August. The blue-winged pitta incubates for about 18 days, the nestlings fledge in 18 to 20 days, and there is only one brood. This is probably standard for the family.

NEW ZEALAND WRENS

The three species alive today in the family Acanthisittidae (or Xenicidae) may be survivors from an ancient colonization of New Zealand. They have a primitive syrinx structure, weak songs, and are weak fliers. Four species were already widespread before humans arrived with their animals. The rifleman *Acanthisitta chloris* and bush wren *Xenicus longipes* were originally common on both North and South Islands of New Zealand. In recent years the bush wren has been recorded at only one North Island locality and is rare in the South. The rock wren *X. gilviventris,* restricted to the South Island, was first described in 1867 and is an inhabitant of Alpine and subalpine scrubland, where it spends much time on the ground searching for insects. Both bush wrens and rock wrens bob when alighting from a short flight. The Stephen Island wren *X. lyalli,* discovered on that island in 1894, was quickly exterminated by the lighthouse keeper's cat.

The rifleman forages for insects on tree trunks like a creeper and is common in beech forests on both islands from sea level to 350 meters (1,150 feet). It breeds from August to January and builds loosely woven nests of moss and plant debris, with a side entrance, in tree hollows or behind loose bark and sometimes on the ground in protected places. After the chicks fledge, the family remains together for several weeks. The bush wren has somewhat similar nesting habits, but builds its nests in low shrubbery from August to December. The rock wren nests among rocks in exposed areas, from September to December.

H. ELLIOTT MCCLURE

◄ A female rifleman visits her nest in a tree cavity. The New Zealand wrens are an isolated group with, apparently, no near relatives; they are unusual among passerines in that females are substantially larger than males, although males are more brightly colored.

M.F. Soper

► *Looking much like old-fashioned clay
ovens, the nests of the rufous hornero
are a common sight along the roadside
fences of the grasslands of South
America.*

OVENBIRDS & THEIR ALLIES

The diverse furnarids of South and Central America comprise more than 244 rust-colored ovenbirds (family Furnariidae), 52 trunk-climbing woodcreepers (Denrocolaptidae), 37 skulking tapaculos (Rhinocryptidae), and 272 brightly patterned antbirds (Formicariidae). With 60 upgraded or new species (in 2 new genera) in the past few years, the eventual total is likely to pass 700 species.

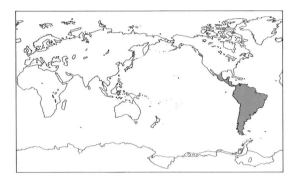

BIRDS OF WOODLAND AND FOREST

The rufous hornero *Furnarius rufus* and some of its relatives are familiar birds near houses, stalking about lawns and singing noisy duets; the barred antshrike *Thamnophilus doliatus* sometimes nests in bushy backyards. Pairs of the recently discovered pink-legged acrobat *Acrobatornis fonsecai* swing unseen from high twigs of shade trees over cocoa plantations in eastern Brazil.

These birds prefer dense foliage, gleaning their prey from leaves, tree trunks, and leaf litter in the undergrowth; because they tend to walk rather than fly, the closer the leaves and accompanying

insects and spiders, the better. Antbirds, and some woodcreepers, are less restricted to dense foliage because they can fly for prey. A few species, such as the jaylike giant antshrike *Batara cinerea* of southern regions, vary the usual diet with small frogs and snakes, eggs, and the nestlings of other birds.

OVENBIRDS

Ovenbirds received their name because the rufous hornero and a few of its relatives build a mud nest like a traditional clay oven, with an entrance at the left or right and the nest chamber inside the remaining half. Most others make oven-shaped nests of twigs or moss, although some hide cup nests inside burrows, crevices, or even in dense foliage. Each family group of the firewood-gatherer *Anumbius annumbi* works together for weeks each year to make two or three wheelbarrow-load nests where all can roost to avoid the cold southern nights. In spring one nest will be occupied by the adult pair for breeding.

The rufous-fronted thornbird *Phacellodomus rufifrons* builds family apartments of sticks at the tips of drooping limbs of savanna trees. The wrenlike rushbird *Phleocryptes melanops* glues wet marsh leaves together to make a cardboard-like oven nest. Closed nests are hard to build, but they do protect against sun, cold, and the birds' predators. Using an oven nest with a narrow entrance poses no problem for these birds, because they are adapted to walk and creep in narrow places when they forage. In contrast, stick nests are not safe from the striped cuckoo *Tapera naevia*, which lays eggs that the unsuspecting hosts will incubate and raise. Some nests attract persistent nest-robbers, notably the oriole *Icterus icterus*, which takes over the nests of thornbirds.

The unusual names of several ovenbirds refer to their behavior or morphology. The campo miner *Geositta poeciloptera* pumps its short tail as it walks over newly burnt Brazilian savannas, and excavates nest holes in armadillo burrows. *Cinclodes* species are sometimes called shaketails because they jerk their tails upward on landing. The wiretail *Sylviorthorhynchus desmursii* has tail filaments more than twice as long as its body, and the softtail

Gunter Ziesler/Bruce Coleman Ltd

Thripophaga macroura has a yellow train. Foliage-gleaners and treehunters (many *Philydor, Automolus,* and *Thripadectes* species) clamber in, or dig into, dense epiphytes or dead leaves, following more alert species that call the alarm if a predator appears. Leaftossers (various *Sclerurus* species) toss or rake leaves on the forest floor. The streamcreeper *Lochmias nematura* tosses leaves to uncover insects along limpid creeks, but a few individuals at sewer outlets above Rio de Janeiro give the whole species the local name "President of Filth". Nuthatch-like bark-probing birds include *Xenops* species in the forests, the recurvebill *Megaxenops parnaguae* in dry woods of northeastern Brazil, and the treerunner *Pygarrhichas albogularis* of woodlands in Patagonia.

We know little about the reproductive behavior of these birds. Male and female build a nest, incubate a few whitish or bluish eggs for 15 to 22 days, and raise nestlings for 13 to 29 days; fledglings receive food for unknown periods and stay with parents for up to a year. One adult buffy tuftedcheek *Pseudocolaptes lawrencei,* studied by Alexander Skutch in Costa Rican cloud forest, built a nest, incubated, and fed its young alone.

WOODCREEPERS

Until the 1930s the family Dendrocolaptidae ("tree hewers") included both woodcreepers and ovenbirds. The former are essentially ovenbirds with strong toes and curve-tipped tails used when climbing tree trunks. Voices of the two groups are similar—long chatters or sharp cries—but only some ovenbirds make a duet. Widely confused with woodpeckers in Latin America, woodcreepers are easily distinguished by their normal feet (woodpeckers have two toes forward and two back) and by their brown or pale-streaked plumage. Unlike ovenbirds, woodcreepers do best in tall-trunked equatorial forests.

Differences between species are greatest in their bills and behavior. Scythebills (*Campylorhamphus* species) use their thin semicircular bills to probe in small holes. The long-billed woodcreeper *Nasica longirostris* of Amazonian riversides, uses its long straight bill to probe in epiphytes growing on horizontal tree limbs. *Drymornis* and *Xiphocolaptes* species have strong and slightly decurved bills, used to rummage among ground litter, rotten logs, and epiphytes growing on vertical trunks. Less-strong bills, some decurved, are a feature of the trunk-climbing *Xiphorhynchus* or *Lepidocolaptes* species. The wedgebill *Glyphorynchus spirurus* wedges insects from scales of bark. Tiny straight bills help *Sittasomus* and *Deconychura* species to peck small insects on tree trunks or capture them in short flights. Stronger straight bills of *Dendrocincla, Dendrocolaptes* and *Hylexetastes* species denote birds that wait and then sally to the ground or to distant trunks or foliage, especially when army ants are flushing prey. The cinnamon-throated woodcreeper *Dendrexetastes rufigula* uses its straight bill in foliage at the tips of branches, pecking apart wasp nests or lunging for prey like a foliage-gleaner. Surface-pecking species can outmaneuver, even chase, the larger but slower sallying species.

Woodcreepers nest in natural tree cavities, or occasionally in wider holes made by woodpeckers, with little nest lining. Male and female incubate one to three white eggs for 15 to 21 days and care for young in the nest for 19 to 24 days, and afterwards for one month or longer—although in *Dendrocincla* species and a few others, the female works alone.

Among the ant-following woodcreepers there are several different behaviors. Big and dominant *Hylexetastes* individuals form pairs and even let their single offspring follow them for a year. Females of the medium-sized *Dendrocolaptes* attack their terrified mates if food is preempted by large

▲ Though unrelated to treecreepers elsewhere in the world, the woodcreepers of South America are similar in their general appearance and in their tree-creeping behavior. Like woodpeckers they use their stiffened tail feathers as a prop to brace themselves against the trunks of trees. This is the wedge-billed woodcreeper.

Luiz Claudio Marigo

A. Greensmith/Ardea London Ltd

▲ *In its general appearance and habits, the Andean tapaculo resembles a wren. It usually frequents dense undergrowth and shrubbery, where it creeps mouse-like with its tail often cocked over its back. It is very difficult to observe, and has a loud penetrating song.*

▼ *Antbirds are so-called because of their habit of following columns of army ants through the forest. Here a white-tufted antbird perches low in typical stance, alert for insects flushed from the ground by the predatory ants.*

birds or faster ones such as antbirds. Small *Dendrocincla* males stay away from their females: either the females seem wary of predators and the increased danger to pairs; or they don't have enough foraging room to let males stay near if dominant competitors are present. The extreme of social isolation is found in the solitary and retiring tyrannine woodcreepers *Dendrocincla tyrannina* of the equatorial Andes, where ant swarms are rare and predators are well hidden in cold and mossy forests; one sex (probably the male) sings for the other like a bowerbird every morning from a notch in a ridge, so that its long, stirring trill is amplified by the sides of the notch over a wide radius.

TAPACULOS

The name "tapaculo" either refers to "cocked tails" (in very risqué Spanish), or to the repeated "tap-a-cu" of the Chilean white-throated tapaculo *Scelorchilus albicollis*. The resemblance of tapaculos to wrens—in size, colors, and methods of rummaging for insects in dense vegetation—is belied by the big feet and spooky repetitive whistles of these non-songbirds. The covered nostrils of the *Rhinocryptidae* species may help foraging in dense debris. They resemble ovenbirds in making covered or burrow nests and in their nearsighted foraging, and like the ovenbirds they rarely follow army ants; however, they also

resemble antbirds in color patterns, and may be the surviving members of an intermediate lineage.

Now mostly restricted to cold forests atop the Andes and in Patagonia and Chile, tapaculos are skulking and wary, hiding in vegetation, but appearing if the observer waits quietly. One of the 18 *Scytalopus* species—birds that tick on and on, or snore like frogs, and sneak like little gray mice in tangled undergrowth—reaches Costa Rica; another lives in the central Brazilian highlands. Once scientists check the birds more carefully and learn their calls, there may be more than 30 species. Four brightly colored antbird-like crescentchests (genus *Melanopareia*) hop through dense savanna grass on both sides of the Andes and in central Brazil. Three thrush-sized turcas and huet-huets (genus *Pteroptochos*) walk low in Chilean–Argentine beech forests and chaparral. The rest of the family is a diverse scatter of possibly ancient forms: the rusty-belted tapaculos *Liosceles thoracicus* on the floor of Amazonian forests; the tinking gray males and rufous females of two tuft-nosed bristlefronts (genus *Merulaxis*) in the litter of eastern Brazilian forests; white-specked rufous ocellated tapaculos *Acropternis orthonyx* in bamboo atop the equatorial Andes; two gallitos ("little roosters", genera *Rhinocrypta* and *Teledromas*) in the semi-deserts of interior South America; and the spotted bamboo-wren *Psiloramphus guttatus,* giving mysterious screech-owl sounds in southern Brazil.

Nesting and social behavior are little known, except for a few burrow or oven nests with white or blue eggs. Small families and pairs are registered in a few species. They hide in foliage rather than join mixed-species flocks.

ANTBIRDS

Antbirds are so-named because some species cling to vertical stems just above the jaws and stings of moving army ant swarms, darting after fleeing insects and other prey. They may be accompanied by woodcreepers, tanagers, ground-cuckoos, and a motley crowd of occasional followers, including lizards. Several antbird species may compete above a single swarm, the larger species taking the center and supplanting the smaller ones peripherally. In compensation, most small species can find enough prey away from ants, or at small swarms. Large species, which get enough food only at the biggest swarms, have become very rare and subject to extinction. One speedy Amazonian species, the small white-tufted antbird *Pithys albifrons,* specializes at darting among larger birds to grab a contested bite. However, it has to fly long distances to find an ant colony with few large birds, and each individual needs an area about 3 kilometers (1¾ miles) in diameter. Specialized birds like this, and the large ant followers, tend to vanish when only small areas of forest are preserved.

The bright blue or red bare faces of several species appear to imitate the eyes of forest cats and

D. Wechsler/Vireo

therefore frighten predators or competitors. Patterns are often bright black and white or rust, but many are gray or black, and females may be a less-colorful brown. White or buff back patches, wing corners, and tail tips, which are usually concealed, may be spread in fights for territories. White-bearded antshrikes *Biatas nigropectus* are colored like the white-collared foliage-gleaners *Philydor fuscus* with which they associate, probably to avoid being singled out if a hawk attacks. Most antbirds fly only short distances in the understory or foliage of the forest; none appear in bright sunlight, and different species often occur on different sides of large rivers in the Amazon region.

Antthrushes (*Chamaeza* and *Formicarius*) walk on the forest floor like bantams, pounding their short upraised tails as they go. Their spooky whistles and white eggs laid in cavities resemble those of tapaculos. Antpittas (*Pittasoma, Grallaria, Hylopezus, Myrmothera,* and small *Grallaricula*), which are big-headed, stub-tailed, and mottled, hop on low vines and the forest floor giving owl-like calls. Nests tend to be low platforms with two greenish eggs. Only the two *Pittasoma* species regularly "bound" like kangaroos around ant swarms. Genetic studies indicate that the 60 species of antthrushes and antpittas may be related to tapaculos and ovenbirds and be the only true members of the Formicariidae, with the remaining 204 antbirds belonging in a more distant family, Thamnophiidae.

The eight species of gnateaters (*Conopophaga*) may also be a separate family related to antthrushes and tapaculos. They are small, short-tailed, brown and black birds, and males often have a white tuft behind the eye. They peck insects nearby or snap them from the forest floor. Nests are bulky cups with two speckled brownish eggs.

The 204 other antbirds, including most ant followers, are among the commonest birds in Amazonian and trans-Andean lowland forest. In second-growth forests, however, few survive. Some species hop or walk on the forest floor, but most peck or sally short distances for prey in the mid-level foliage, while a few inhabit the subcanopy.

Antwrens (*Myrmotherula*, plus other genera) are small, short-tailed, and warblerlike. Different species have separate niches in the Amazon: commonly one lives near the ground, three or more in the understory (at least one investigating dead leaves, and one in dense tangles), one or more in the canopy, plus others in streamside or forest-edge. Often they join understory flocks of up to 50 different species of birds, including woodcreepers and ovenbirds. In each flock may be one or more slightly larger antvireos (*Dysithamnus*) and various thick-billed antshrikes (*Thamnophilus, Taraba,* and *Sakesphorus*), from sparrow to jay size, though these often live apart in thickets. Occasionally with the flocks may be various "antbirds", including ant followers, plus long-tailed tangle living *Cercomacra* or short-tailed undergrowth-haunting *Myrmeciza*.

Luiz Claudio Marigo

▲ *A little-known bird of forest undergrowth, the black-spotted bare-eye is distributed east of the Andes from Colombia to Bolivia.*

The most unusual species are little gray or brown antcatchers (*Thamnomanes*), which are actually good flycatchers, perching upright and snapping up insects flushed by the mixed flock. Like drongos in Old World flocks, they faithfully sound the alarm if a hawk or other danger appears, and for this service they attract others. (Sometimes, however, they give a false alarm and catch insects in the confusion!) Amazonian Indians, and later Brazilians, thought the "uirapurus" very potent medicine because lots of birds followed them.

A mixed flock allows rummaging specialists like ovenbirds to seek food while alert *Thamnonmanes* antcatchers keep a lookout for predators. Ant-catchers need a large territory, however, and when trees are logged leaving only small areas of forest, the resulting loss of *Thamnomanes* species can lead to the predation and loss of other species.

Some antbirds, notably antwrens of the genus *Formicivora*, live in shady scrub in grasslands, on rocky slopes, or on beach dunes. One, the marsh antwren *Stymphalornis acutirostris*, was recently discovered in beach marshes in southeastern Brazil.

The songs of antbirds tend to be series of whistles, with rough sounds for enemies and soft ones for family. When courting and before copulation, the male feeds the female. Both build the nest, a small cup or, less often, an oven shape. At night the female incubates or broods, but both sexes incubate the two (rarely more) spotted whitish eggs for 15 to 20 days, then care for nestlings for 9 to 18 days and for one fledgling each for a month or so. The association of juvenile birds with parents lasts for between one month and a year.

EDWIN O. WILLIS

TYRANT FLYCATCHERS & THEIR ALLIES

Short-tailed pygmy-tyrant
Myiornis ecaudatus
Total length: 6 cm (2⅖ in)
Weight: 5 g (⅕ oz)

Amazonian umbrellabird
Cephalopterus ornatus
Total length: 45 cm (17¾ in)
Weight: 400 g (14 oz)

CONSERVATION WATCH
!!! The kinglet calyptura
Calyptura cristata and Peruvian
plantcutter *Phytotoma raimondii*
are critically endangered.
!! 8 species are listed as
endangered: banded cotinga
Cotinga maculata; ash-breasted
tit-tyrant *Anairetes alpinus*;
Kaempfer's tody-tyrant
Hemitriccus kaempferi; Atlantic
royal flycatcher *Onychorhynchus
swainsoni*; Alagoas tyrannulet
Phylloscartes ceciliae; Antioquia
bristle-tyrant *Phylloscartes lanyoni*;
Minas Gerais tyrannulet
Phylloscartes roquettei; giant
kingbird *Tyrannus cubensis*.
! 17 species are listed as
vulnerable.

T he suborder Tyranni is restricted to the Americas, where it is widely distributed in virtually all habitats, elevations, and latitudes. There is great variability in size, plumage coloration, and behavior within this group, and the extreme diversity found both within and between families makes it very difficult to generalize about them. In fact, scientists remain uncertain as to which family some species belong. These diverse birds are united in this suborder by a particular arrangement of the muscles of the syrinx or vocal organ.

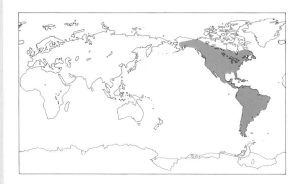

TYRANT FLYCATCHERS
The family Tyrannidae is found throughout North, Central, and South America. It includes not only the tyrant flycatchers but also birds commonly

known as phoebes, elaenias, kingbirds, flatbills, and wood-peewees. The sexes look alike, generally with plumage that is a mixture of greens, browns, yellows, and white. In most species a strong pair-bond is formed for the duration of the breeding season, and both sexes help to raise the young. Of the nearly 400 species in 114 genera that have been identified so far, the greatest diversity is found in the New World tropics. Each species exploits the environment in a slightly different way, and so intricate is this division of resources that some avian communities in South America are known to contain in excess of 60 flycatcher species.

The difference in food preferences is one reason why multiple species can coexist. Obviously, fruit-eating and insect-eating flycatchers can live

► *The scissor-tailed flycatcher is a common and conspicuous bird of open country from the south-central United States to Panama. Noisy and pugnacious, it frequently harasses hawks and other birds much larger than itself.*

R.C. Simpson/Tom Stack & Associates

together without direct competition for food. However, the great majority of flycatchers, as their name suggests, eat insects. How can so many similar species eat the same type of food and still be found together? The answer is that each species employs a slightly different combination of prey size, habitat, vegetation type, foraging position, and capture technique. Scientists believe that it is this ability to finely divide the available resources, coupled with the fact that the New World tropics provide birds with rich and diverse resources to exploit, that gave rise to the largest of passerine families. (Their counterparts in the Old World are the flycatchers in the family Muscicapidae.)

MANAKINS

Manakins are small, brightly colored fruit-eaters with short tails and short, wide bills. There are 57 species (placed in 19 genera in the family Pipridae) confined to lowland forests of South and Central America. The females look remarkably similar between species, most with green or olive plumage. In contrast the males are generally brightly colored, usually with bright blues, reds, or yellows on a black background.

This high degree of color difference between the sexes is indicative of a complex social behavior. Male manakins congregate at traditional places in the forest, known as "leks", to attract mates. In some species, males clear dance arenas on the forest floor; in others, males prepare dance perches

Luiz Claudio Marigo

▲ In a bizarre courtship display the male wire-tailed manakin (left) rapidly brushes the female's throat with the wire-like tips of his tail feathers. The great kiskadee (center) is one of the more conspicuous tyrant flycatchers of tropical America, often perching on roadside fenceposts and telephone wires. Only the male Guiana cock-of-the-rock (right) is brightly colored; the female is plain, dowdy, and gray. The common name hints at two features: males congregate at leks to display like cocks; and females come to the lek to mate, then retire alone to build their nests in rock crevices.

◄ Manakins are notable for their extraordinary group displays and complex courtship behavior, but many species remain unstudied. This species, the helmeted manakin, is a little-known bird that inhabits the tablelands of central and southern Brazil.

▶ The plantcutters, a family of only three species of exclusively South American distribution, are among the very few purely vegetarian passerines. They inhabit open country and closely resemble finches in general appearance and behavior. This is the rufous-tailed plantcutter.

J. Dunning/Vireo

by stripping branches of leaves. Females are attracted to these leks by the calls of the males. When a female arrives the resident males perform their elaborate courtship displays which highlight their bright plumage. The displays are stereotyped within a species and can include vocalizations, wing snaps, short flights, hops, side-steps, and rapid tail movements. If the female is receptive, the pair will mate at the conclusion of the display. The display and mating constitute the extent of the pair-bond in manakins; the female, having previously built a nest, will lay eggs, incubate, and feed the young without further assistance from the male. In one genus, *Chiroxiphia,* the reproductive display requires cooperation between several males, only one of whom will have the opportunity to mate with the female.

COTINGAS
Members of the family Cotingidae (61 species in 25 genera) are fruit-eating birds, generally weighing 50 to 100 grams (1¾ to 3½ ounces). They are restricted to Central and South America, occur in low densities, and usually frequent the upper portions of forest canopies. These last two observations lead directly to the remaining generalization, which is that cotingas are very poorly known. In some genera (for example, *Lipaugus*) male and female plumage is similar to that of flycatchers; the sexes look alike, and have drab green and brown plumage. In most other genera, however, the males are brightly colored and only the females possess the more cryptic coloration. Complex social behavior appears to be common within this family, but only the cocks-of-the-rock (genus *Rupicola*) have been well studied. There is even a suggestion that, like the manakin genus *Chiroxiphia,* the reproductive displays of some cotinga species require cooperation between males.

When birds specialize on fruit as a food source, as is true for the cotingas, there is ample opportunity for plant and bird to become strongly interdependent. For example, the white-cheeked cotinga *Ampelion stresemanni* of the Peruvian Andes feeds almost exclusively on mistletoe fruit. After a meal it regurgitates the seeds onto branches, where in time the seeds will germinate and grow. Clearly the mistletoe is dependent on the cotinga for seed dispersal because there is no other fruit-eater at this elevation. Therefore, without the cotinga the seeds would simply fall to the ground, where they cannot survive.

SHARPBILL
Named for its bill shape, the sharpbill *Oxyruncus cristatus* is a canopy-dwelling fruit-eating bird about which little is known. Its range is probably between Costa Rica and southern Brazil, but it has been found at few localities. Attempts to determine whether sharpbills are more closely related to flycatchers or to cotingas have been equivocal. Scientists can only conclude that this species is a survivor of a very old lineage within the Tyranni.

PLANTCUTTERS
The plantcutters, three species in the genus *Phytotoma,* are confined to dry habitats in southern South America. These species have the distinction of being the only passerine birds known to depend on leaves and fleshy stems for the bulk of their diet. They have serrations on the edge of the bill that enable them to cut leaves and stems into pieces small enough to eat. Biochemical studies indicate that they are most closely related to the cotingas.

SCOTT M. LANYON

LYREBIRDS & SCRUB-BIRDS

KEY FACTS

ORDER PASSERIFORMES
SUBORDER OSCINES
FAMILY MENURIDAE
• 1 genus • 2 species
FAMILY ATRICHORNITHIDAE
• 1 genus • 2 species

SMALLEST & LARGEST

Rufous scrub-bird *Atrichornis rufescens*
Total length: 16–18 cm (6½–7 in)
Weight: 30 g (1 oz)

Superb lyrebird *Menura novaehollandiae*
Total length: 80–100 cm (31½–39½ in)
Tail length: 50–60 cm (20–23½ in)
Weight: up to 1.2 kg (2½ lb)

CONSERVATION WATCH

! The noisy scrub-bird *Atrichornis clamosus* and rufous scrub-bird *Atrichornis rufescens* are listed as vulnerable.

L yrebirds and scrub-birds are surviving members of an ancient and diverse radiation of Australian songbirds, so while quite distinct, the two families are each other's closest relative. All species are terrestrial, have weakly developed powers of flight, and live in dense vegetation. Despite the difference in size they have a number of ecological similarities because of common adaptations to a similar habitat.

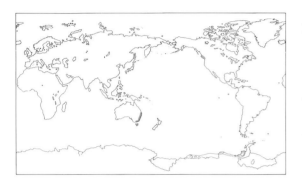

LYREBIRDS

The suberb lyrebird *Menura novaehollandiae* occurs in a narrow belt of Australia from northeastern New South Wales to southeastern Victoria. It was introduced into Tasmania in 1934. Its habitat is wet eucalypt forest and temperate rainforest.

Albert's lyrebird *M. alberti* is restricted to a small belt of subtropical forest from northeastern New South Wales to extreme southeastern Queensland.

Lyrebirds are pheasant-sized, with brown to rufous plumage, long powerful legs, short rounded wings, and a long tail with modified feathers; the outer two feathers in the superb lyrebird are shaped like a Greek lyre, hence the name. They are elusive and difficult to observe in their dense habitat, where they are fast, agile runners when danger appears. They rarely perch in trees except to roost, ascending by jumping from branch to branch, and descending in the morning by gliding. The diet is mainly invertebrates which they expose by digging, ripping apart rotten logs, or turning over stones with their powerful feet.

Their most conspicuous characteristic is their loud territorial song, 80 percent of which may be

Hans & Judy Beste/Auscape International

◀ Largest of all songbirds (passerines), the superb lyrebird feeds on insects and other invertebrates living on the forest floor, searching the leaf litter for them with methodical, raking sweeps of its large clawed feet. Smaller birds such as yellow-throated scrubwrens frequently attend it in a "crumbs from the lord's table" relationship, feeding on the smaller insects overlooked or ignored by the larger bird but disturbed by its raking.

▶ *The noisy scrub-bird is aptly named: notorious among bird-watchers for being so elusive and so agile in dense cover that it is all but impossible even to glimpse, let alone observe, it is also notable for the almost ear-splitting intensity of its calls.*

Graeme Chapman

▼ *Like the noisy scrub-bird in behavior and general appearance, the rufous scrub-bird can barely fly. It lives close to the ground in dense cover, slipping mouse-like through the leaf-litter. In display the male cocks his tail, droops his wings, fluffs his chest feathers, and delivers his extraordinarily penetrating song. So inflexible are the bird's habitat requirements and so fragmented its distribution that its total numbers may not exceed 1,000, and its status is extremely precarious.*

Glen Threlfo/Auscape International

mimicry of other birds—occasionally they may mimic barking dogs. The adult male superb lyrebird establishes a territory of 2.5 to 3.5 hectares (6 to 9 acres) when sexually mature; the female's nesting territory may be within or overlapping the male territories. Males defend their territories, especially during the winter breeding season, by chasing intruders, singing, or displaying on earth mounds (well-concealed platforms of vines and fallen branches for Albert's lyrebird), which they have constructed throughout their territories. The displays are spectacular; the tail is fanned and thrown forward over the head and vibrated while the bird dances and sings.

The male mates with a number of females attracted to his displays. The female builds a domed nest, usually less than 2 meters (6½ feet) above ground, incubates the single egg for 47 days, and feeds the nestling until it leaves the nest when about 50 days old. The young bird stays with its mother for about eight months after fledging. After the breeding season the birds are more mobile and wander in small groups through the forest.

The above is based on observations of the superb lyrebird. The little that is known about Albert's lyrebird suggests that there are few major differences in overall biology between the two species.

SCRUB-BIRDS

The noisy scrub-bird *Atrichornis clamosus* originally survived in only one small locality on the south coast of Western Australia. Management to prevent fire and a translocation program to a nearby area has resulted in a 15-fold increase in numbers. Its main habitat is the boundary between swamp and forest and in wet gullies. The rufous scrub-bird *A. rufescens* has a discontinuous distribution on the coast around the Queensland–New South Wales border, in wet temperate forest and subtropical rainforest.

The habitat of both species is dense, so scrub-birds are rarely seen. They are small, solidly built brown birds with long legs and tail, and short rounded wings; they rarely fly, and then only for a few meters. The males have a loud territorial song (the females give only alarm notes). The rufous scrub-bird male also uses mimicry, and the noisy scrub-bird has another quieter song which uses segments of other birds' songs. Noisy scrub-bird males have territories of 5 to 10 hectares (12½ to 25 acres) within which the females have nesting territories. The nesting behavior is similar to that of lyrebirds, but of shorter duration, and the rufous scrub-bird breeds in spring/summer. Both species maintain their territories throughout the year.

G. T. SMITH

LARKS & WAGTAILS

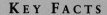

KEY FACTS

ORDER PASSERIFORMES
SUBORDER OSCINES
FAMILY ALAUDIDAE
• *c.* 15 genera • *c.* 79 species
FAMILY MOTACILLIDAE
• *c.* 5 genera • *c.* 60 species

SIZE

Larks (Alaudidae)
Total length: 11–19 cm (4¾–7½ in)
Weight: 13–45 g (½–1½ oz)

Wagtails (Motacillidae)
Total length: 14–17 cm (5½–6⅔ in)
Weight: 13–32 g (½–1 oz)

CONSERVATION WATCH
!!! Rudd's lark *Heteromirafra ruddi* is critically endangered.
!! There are 5 species listed as endangered: Raso lark *Alauda razae*; Archer's lark *Heteromirafra archeri*; Sidamo lark *Heteromirafra sidamoensis*; Ash's lark *Mirafra ashi*; ochre-breasted pipit *Anthus nattereri*.
! The 5 vulnerable species are: red lark *Certhilauda burra*; Degodi lark *Mirafra degodiensis*; Botha's lark *Spizocorys fringillaris*; yellow-breasted pipit *Anthus chloris*; Sokoke pipit *Anthus sokokensis*.

These two families, the Alaudidae and the Motacillidae, are small ground-dwelling birds, represented worldwide except in extreme latitudes and oceanic islands. Larks are of course known for the beautiful song of many species. Some of the pipits (which are in the same family as wagtails) also have a lovely musical call in flight and superficially resemble larks.

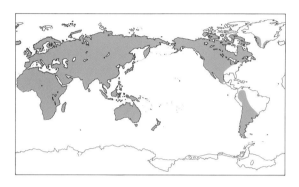

LARKS

Not all larks have the vocal powers of the renowned skylark *Alauda arvensis* of Eurasia, though many others sing well during the course of their aerial song-flights. Some sing from tree-stumps or posts or even anthills, and others produce clapping or fluttering sounds through wing-action during their nuptial displays. Because of its popularity, the skylark has been introduced to Vancouver Island on Canada's Pacific coast (it has spread to the nearby San Juan Islands of Washington State), New Zealand, and Australia (which already had one other species, the singing bushlark *Mirafra javanica*).

The lark family (Alaudidae) is centered mainly on temperate regions of the Old World in generally open habitats. It comprises about 79 species, with the greatest concentration in Africa. The most numerous is the shore or horned lark *Eremophila alpestris*, which has the widest distribution and occupies a great variety of habitats from Arctic tundra to temperate grasslands and even desert. Other species are usually more particular about their choice of terrain.

Richard T. Mills

◄ *A pair of skylarks raise up to four broods per season. This bird may be feeding an early brood: as in many well-studied birds, the average clutch size has been found to vary not only geographically but also through the season, rising from just over three in early spring to four in summer.*

▶ Slender and graceful in build, wagtails are named for their habit of persistently waving their tails up and down. The gray wagtail is seldom found away from water, especially rushing mountain streams and swiftly flowing rivers.

Laurie Campbell/NHPA

Typical larks are generally streaked brown over the upper parts, wings and tail, and white or buff on the underside, the breast usually streaked with dark brown. Some have a small crest on the head, and some show white over the lateral tail. In most, the bill is slender and slightly decurved, but in others it is robust. The legs and toes are long; in species that inhabit grassland the claw of the hind toe is extended. All walk rather than hop.

Larks feed on insects and other invertebrates as well as seeds and grain. They build simple cup-shaped nests on the ground; these are sometimes completely exposed and sometimes sheltered at the base of a tuft of grass or under a low bush. In desert a partial canopy may be added to shield the incubating female from the heat of the sun. Many larks exhibit plumage colors that match the color of the soil on which they breed, resulting in such forms within a species being classified as different subspecies or races. The horned lark, for example, has many differently colored forms—breeding as it

does in places as varied as Franz Josef Land and Novaya Zemlya in northern Eurasia, southern Mexico, and the savannas of Colombia.

Finchlarks

The finchlarks (genus *Eremopterix*) are small, the sexes differ markedly from one another, and their habitat is mainly semi-desert country. One species is restricted to India, Pakistan, and Sri Lanka; all others live in Africa. They are more variegated in their color patterns than true larks and, as their vernacular name implies, have short conical finch-like bills. Like other larks they are gregarious when not breeding, and they are nomadic when food becomes seasonally scarce in their severe habitat.

WAGTAILS AND PIPITS

The 60 or so species in the family Motacillidae are small and of slender build, some of them longish-tailed. They are mainly ground-nesters and almost entirely insectivorous.

Wagtails

Typical wagtails are characteristic birds near running water on riverbanks and in moist grassland. Four species are resident in southern and eastern Africa and Madagascar (they do not migrate), whereas the wagtails that breed in northern countries are highly migratory. In these northern forms a breeding plumage is assumed by the adult males, so although wagtails are generally classified as 11 species, there is extensive racial variation—the seasonal change of dress being most markedly illustrated in the yellow wagtail *Motacilla flava,* of which 17 subspecies are recognized. The yellow wagtail is the only one that breeds in North America, in the northwest of Canada and Alaska. This species and the yellow-headed wagtail *M. citreola* are inhabitants of Eurasian steppes and pasture where they feed among grazing cattle. Of the other *Motacilla* species, four are white or pied; the purely African *M. capensis* and its Madagascan counterpart *M. flaviventris* are plainer, and two are long-tailed and choose purely riverine habitats (*M. cinerea* and *M. clara*).

An atypical species, the eastern Eurasian forest wagtail *Dendronanthus indicus,* is restricted to forest on its breeding grounds.

Pipits

The longclaws (genus *Macronyx*) represent the largest and most colorful of the pipits. All seven species live in Africa, centered on the eastern and southern savannas. The largest are two yellow-throated species, *M. croceus* and *M. fuelleborni,* and the orange-throated Cape longclaw *M. capensis;* the pink-throated *M. ameliae* and *M. grimwoodi* are somewhat smaller; and the remaining two are small and yellow-throated, *M. flavicollis* in the Ethiopian highlands, and *M. aurantiigula* in East Africa. These birds demonstrate an interesting evolutionary convergence with the meadowlarks found in the Americas.

Linking the decorative longclaws of Africa and the dull-colored pipits of the genus *Anthus* are, in Africa, the golden pipit *Tmetothylacus tenellus,* plus a couple of anomalous forms, the yellow-breasted pipit *Hemimacronyx chloris* in the Drakensberg Mountains of southeastern Africa, and the Kenyan yellow-breasted pipit *H. sharpei* (also known as Sharpe's longclaw) found in the highland grasslands of Kenya.

The *Anthus* pipits are lark-like birds of open country and montane environments, generally brownish above, plain or moderately streaked, light buff on the underside, with the breast and sides usually streaked with dark brown. In many, the outer tail-feathers are broadly marked with white, but because the differences between species are subtle and the plumage is altered through bleaching and wear, field determination can be difficult. Northern species of pipits are highly migratory, and even in tropical species there is much post-breeding movement associated with local seasonal declines in the availability of food. The *Anthus* pipits probably originated in the Old World temperate regions, from which they radiated extensively: 14 species are seen as basically Palaearctic (Eurasian), 13 endemic to Africa, four to Indo-Malaysia and Australasia, eight to Central and South America, and two to North America. Why the family should be so poorly represented in North America is currently inexplicable.

P. A. CLANCEY

◀ A European meadow pipit. Pipits are very ordinary in appearance but they are among the most widespread of all songbirds. Though the group reaches its greatest diversity in Eurasia, one form or another occurs in open country of all kinds almost everywhere except Antarctica; one species even inhabits the remote and inhospitable island of South Georgia.

Richard T. Mills

SWALLOWS

KEY FACTS

ORDER PASSERIIFORMES
SUBORDER OSCINES
FAMILY HIRUNDINIDAE
• c. 20 genera • c. 82 species

SIZE

Total length: 12–25 cm (4¾–10 in)
Weight: 10–45 g (½–1½ oz)

CONSERVATION WATCH
!!! The white-eyed river martin *Pseudochelidon sirintarae* is critically endangered.
! The vulnerable species are: blue swallow *Hirundo atrocaerulea*; white-tailed swallow *Hirundo megaensis*; Red Sea swallow *Hirundo perdita*.

The family Hirundinidae includes about 82 swallows and martins which are aerial-feeding insect-eaters distributed throughout the world's temperate and tropical zones (except some islands), and two species of river martins.

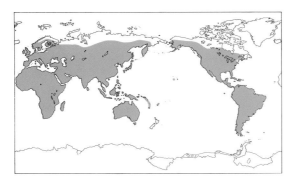

SWALLOWS AND MARTINS

The plumage of typical swallows (genus *Hirundo*) is glossy blue-black on the upper parts, dark on the wings and tail, the latter spotted subterminally with white, and the outer tail-quills in adults extended into narrow filaments; many species have patches of red-brown over the forehead or fore-throat. Others may have the tail squared and lack the streamers of the *Hirundo* species; the top of the head and lower back are tawny or red-brown and the underside is streaked (as in *Cecropis* and *Petrochelidon* species). Crag martins, which are nowadays classified in the genus *Hirundo* (formerly *Ptyonoprogne*), are dull-colored birds of dry and even desert regions in the Old World. The house martin *Delichon urbica*, and its allies the Asiatic house martin *D. dasypus* and the Nepal house martin *D. nipalensis* distributed across Europe and Asia, are glossy blue-black above and have short legs and toes covered with feathers as a protection

▶ *The barn swallow usually builds its nest in some human-built structure such as a bridge, shed, or barn, but it shows a marked preference for buildings in which cattle or other domestic stock are kept. Very occasionally it nests in ancestral sites such as crevices in cliffs. It breeds across much of the Northern Hemisphere, wintering in South America, Africa, India, and Southeast Asia.*

Wisniewski/Zefa

Richard T. Mills

◄ Sand martins are strongly gregarious and nest in colonies, excavating holes in sandy river banks and quarries. They show strong fidelity to the site, returning every season but usually drilling a fresh burrow each time. The species is widespread across North America and Eurasia, but the European population suffered several abrupt declines during the period 1960–1990: one colony in Scotland collapsed from 900 pairs in 1982 to fewer than 200 pairs two years later. This decline has been linked with a concurrent series of devastating droughts affecting the region immediately south of the Sahara, from Eritrea west through the Sudan to the Sahel, where most of the birds spend the winter.

against the low temperatures they encounter. Other martins are robust, glossy American species of the genus *Progne,* and the sand martins (genus *Riparia*) which are small and dull-colored.

These birds feed on insects taken during flight, and some may be found consorting when insects are temporarily locally abundant. They may also be seen alongside swifts when insect swarms are present near to the ground before a storm. Mainly silent birds, they give voice to twittering and short warbling songs delivered while on the wing or perched on bare twig. Most are strong migrants, especially the species of *Hirundo, Delichon, Riparia* and others, some of which move north to breed at high latitudes. Even tropical species have their seasonal movements—for example, the Malagasy Mascarene martin *Phedina borbonica* ranges to eastern Africa after breeding on Madagascar. In the Afrotropics the roughwings (genus *Psalidoprocne*) are usually quite sedentary.

Most swallows build solid nests in the shape of half-bowls, saucers, and even retorts, fashioned out of mud pellets and straw, the eggs resting in a cup of fine grasses, hair, and curly feathers. Species of the genus *Riparia* nest in colonies, tunneling into vertical sandy banks and creating a nest of grasses and feathers at the end of the tunnel.

RIVER MARTINS
Both species of river martin are red-billed, short-tailed aberrant swallows. A native of the Zaire river system of Africa, the African river martin *Pseudochelidon eurystomina* regulates its breeding cycle to the drop in the level of a major river so that it can breed (in colonies) in burrows on the exposed sandy bars. The white-eyed river martin *P. sirintarae* was discovered in Thailand and named only in 1968; its breeding grounds are probably the middle reaches of the Mekong River.

P. A. CLANCEY

CUCKOOSHRIKES

SMALLEST & LARGEST

West African wattled
cuckooshrike *Campephaga lobata*
Total length: 18 cm (7 in)
Weight: 20 g (⁷⁄₁₀ oz)

Ground cuckooshrike
Pteropodocys maxima
Total length: 28 cm (11 in)
Weight: 111 g (4 oz)

CONSERVATION WATCH

!! The Réunion cuckooshrike
Coracina newtoni is listed as
endangered.
! There are 6 vulnerable species:
western wattled cuckooshrike
Campephaga lobata; Buru
cuckooshrike *Coracina fortis*;
McGregor's cuckooshrike
Coracina mcgregori; black-bibbed
cicadabird *Coracina mindanensis*;
white-winged cuckooshrike
Coracina ostenta; Mauritius
cuckooshrike *Coracina typica*.

C uckooshrikes are so-called because of their bustle of hard-shafted yet soft and
loosely attached rump-feathers which is similar to that of Old World cuckoos.
The family, Campephagidae, is confined to the tropics of Africa south of the
Sahara and also from Afghanistan and the Himalayas, east across Asia to the Japanese
islands, and south to Southeast Asia, Australia, and the Pacific islands.

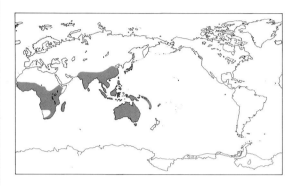

GRAY TREE-DWELLERS

Cuckooshrikes in the genus *Coracina* are
moderately large, gray or blackish and white, and
generally barred. Male and female are similar in
appearance. Their bill is relatively heavy, broadly
based, and hook-tipped, and they have short legs.
The 43 species of this genus and its immediate
allies are centered numerically on Southeast Asia
and Australia, and with some species having a

remarkable variety of forms—no less than
33 subspecies of the cicadabird *C. tenuirostris* are
recognized. In the wholly African genus
Campephaga, comprising five species, the adults are
markedly different: the males are an attractive
glossy blue-black, with some exhibiting bright
patches of yellow, orange or deep red on the wings;
the females are dark olive-brown, barred and
streaked in yellow, yet whitish below.

Cuckooshrikes are of secretive disposition,
hiding among screening foliage as they forage for
insects and fruit, but some *Coracina* species are
more conspicuous and during the non-breeding
season may be seen in small parties. Nests are
usually constructed high up in trees and skillfully
blended into the moss and lichens of tree limbs to
make detection difficult.

The trillers (genus *Lalage*) reach Australia and
the Pacific islands, but the 13 species are mostly
distributed on the mainland and islands of
Southeast Asia. Their plumage is blackish-gray and
white, with some tinged a rust color.

The woodshrikes (genera *Hemipus* and
Tephrodornis), close relatives of the trillers, are four
robust species, gray, black, and white in color, or
else pied, and dwell in the forests and woodlands
of Southeast Asia and Indonesia.

The 10 species of minivets (genus *Pericrocotus*)
are restricted to the east of the family's range,
extending from the Amur River near the
Russia–China border and the islands of Japan,
east to Afghanistan, India, and south to the
islands of Indonesia. Males of many species are
strikingly colored in bright red, orange, and
yellow, although some are by contrast dull-
colored, and all females are less colorful than the
males. The biology of this group appears to differ
little from that of the cuckooshrikes.

P. A. CLANCEY

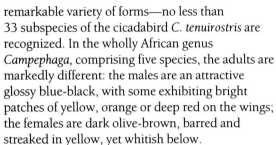

◄ *The barred or yellow-eyed
cuckooshrike inhabits eastern Australia
and the New Guinea region. It is a
nomadic rainforest dweller that feeds
mainly on fruit.*

BULBULS & LEAFBIRDS

These birds are in the main solitary, but readily congregate when food is freely available. The colorful and attractive leafbirds of the family Irenidae occur alongside bulbuls in forested areas of southern Asia.

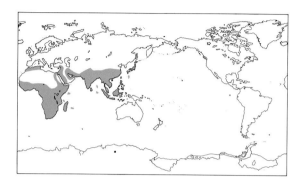

BULBULS

Bulbuls are a mainly tropical Old World family, Pycnonotidae, comprising 15 genera with about 120 species, which are distributed in Africa and from the Middle East across to Japan and south to Indonesia. Significantly they do not reach New Guinea or Australia—although the red-whiskered bulbul *Pycnonotus jocosus* has been introduced (successfully) to Sydney and Melbourne, and also to Florida in the USA. Bulbuls are small to medium-sized, and somberly colored in olive and brown, often whitish or yellow on the underside, with some exhibiting distinctive yellow or red undertail coverts. Many are crested and have hairlike filo-plumes over the back of the head. The bill is relatively robust, often notched towards the tip, and the legs and toes are strong.

Most bulbuls are insect- and fruit-eaters. When certain favored trees are bearing fruit, some species congregate in numbers, but otherwise they tend to be solitary and keep much to screening vegetation, especially in the canopy of high forest and woodlands. However, some exploit the understory and are partly terrestrial—especially so in the moderately gregarious species of *Phyllastrephus* in the African tropics. Several make their presence

KEY FACTS

ORDER PASSERIFORMES
SUBORDER OSCINES
FAMILY PYCNONOTIDAE
• 15 genera • *c.* 120 species
FAMILY IRENIDAE
• 3 genera • 14 species

SMALLEST & LARGEST

BULBULS (PYNCONOTIDAE)

Slender bulbul *Phyllastrephus debilis*
Total length: 14 cm (5 in)
Weight: 13–16.5 g (½ oz)

Yellow-spotted nicator *Nicator gularis*
Total length: 23 cm (9 in)
Weight: 63 g (2 oz)

FAIRY BLUEBIRDS, LEAFBIRDS & IORAS (IRENIDAE)

Common iora *Aegithina tiphia*
Total length: 12 cm (4¾ in)
Weight: 10 g (⅖ oz)

Fairy-bluebird *Irena puella*
Total length: 27 cm (10½ in)
Weight: 70 g (2½ oz)

CONSERVATION WATCH
!!! The Liberian greenbul *Phyllastrephus leucolepis* is critically endangered.
!! The streak-breasted bulbul *Ixos siquijorensis*, dusky greenbul *Phyllastrephus tenebrosus*, and Philippine leafbird *Chloropsis flavipennis* are listed as endangered.
! There are 9 vulnerable species.

◄ The red-vented bulbul (far left) is a common inhabitant of parks and gardens in Southeast Asia. Its calls are cheerful but undistinguished, whereas the golden-fronted leafbird (near left) is notable for its flawless mimicry of the calls of other birds.

Morten Strange

▲ *Displaying the perky self-confident air so characteristic of its family, the yellow-vented bulbul is abundant in lowland forests and clearings throughout much of Southeast Asia. It congregates at night to roost in flocks.*

known in gardens in urbanized areas.

Bulbuls build loosely constructed nests of local plant materials, placing them where they will be concealed in trees, bushes, and creepers. Vocalization generally consists of ringing calls, often in answer to one another from different points of the forest; and when breeding the birds make short warbling and sometimes trilling songs—for example, the somber bulbul *Andropadus importunus* of southern and eastern Africa.

Most of the species are sedentary, but a few from the high northern latitudes are recorded as undertaking post-breeding migratory movements. The most widely distributed genus in the family is *Pycnonotus,* crested birds with colored undertail coverts; the common bulbul *P. barbatus* reaches northwest Africa and the eastern shores of the Mediterranean Sea.

In the African tropics the forest-dwelling species of *Andropadus* and its immediate allies are confusingly similar, and the validity of a couple of

these remains to be resolved. Among these African bulbuls significant deviations from the norm are presented by the golden bulbul *Calyptocichla serina* and the honeyguide bulbuls *Baeopogon indicator* and *B. clamans*: the honeyguide bulbuls mimic the appearance of the true honeyguides, larger birds of the genus *Indicator,* family Indicatoridae, which are seemingly distasteful to predators.

The main concentration of species is in the equatorial rainforests of Africa and similar vegetation in Southeast Asia, the Philippines, and the Indonesian islands. The Asian genus *Hypsipetes* has an extensive range, several species reaching islands in the western Indian Ocean such as the Seychelles, Mauritius, Reunion, Madagascar, and even Comoros, but failing to gain a foothold on the African continent.

FAIRY-BLUEBIRDS, LEAFBIRDS, AND IORAS

An essentially Indo-Malaysian group of colorful arboreal birds, the family Irenidae extends from the Himalayas, the Indian peninsula and Sri Lanka eastwards to southern China, and south to the Philippines and Indonesia. Most species are largely sedentary, but some are on record as undertaking seasonal post-breeding migratory movements.

Fairy-bluebirds

The two fairy-bluebirds *Irena puella* and *I. cyanogaster* are the largest members of the family Irenidae. They are striking birds, the males shining ultramarine blue over the upper parts and black below, the females a duller, greener blue, and with black lores (the area extending from the eyes to the bill-base); all have red eyes. The bill is relatively stout with the culmen (the dorsal ridge from tip to forehead) arched, and the upper mandible notched short of the tip. Fairy-bluebirds feed on fruits from forest trees and shrubs, as well as flower nectar and small invertebrates, foraging in parties of up to eight individuals in the canopy of the monsoon forest. They are seemingly less aggressive towards competitors than their smaller relatives, the leafbirds, and are frequently seen consorting with other fruit-eating birds attracted to trees laden with fruit. Fairy-bluebirds nest in the upper story of forest trees, and their calls have been described as a "twing twing". *Irena cyanogaster* is confined to the Philippines, while the range of *I. puella* extends from there to northwestern India.

Leafbirds

The eight species of leafbirds (genus *Chloropsis*) are largely green; the males have a blue-black band on the breast, marked with vivid blue (as in the blue-whiskered leafbird *C. cyanopogon*); the forehead is green, yellow, or red; some are orange below, and in the blue-winged leafbird *C. cochinchinensis* the wings are blue. The females are duller than the males and lack the dark breast-band. They all have a more slender bill than the bluebirds, which is similarly notched on the upper mandible.

Leafbirds inhabit forests of various types, even mangroves, as well as gardens and parks. Despite their short, rounded wings, their flight is both sustained and swift. Field researchers describe the birds as highly pugnacious, driving away competitors as they forage for fruits, invertebrates, and flower nectar. Keeping as they do to the screening cover of the canopy and densely leaved trees, they are often difficult to see because of their dominant color of green, although while they are foraging, leafbirds are as lively and dexterous as tits of the family Paridae, hanging upside down and assuming various acrobatic postures as they search for elusive prey and food items.

Nests are described as a loose cup, semi-pensile in form, built of local plant materials felted together, and the eggs are pale buff, marked with spots and blotches of red-brown.

IORAS

Ioras (four species in the genus *Aegithina*) are distributed from northwestern India to Borneo and the Philippines. They are the size of small sparrows, mostly green, with the wings starkly black crossed by two white bars. As in other members of the family, the upper mandible of their bill is notched towards the tip; but unlike the bills of *Irena* and *Chloropsis* the culmen is relatively straight and not arched or decurved. The great iora *A. lafresnayei* and green iora *A. viridissima* have an interesting mannerism: they fluff up their long lightly colored flank feathers over their back to give the impression that the rump is white, which it is not. Ioras are fruit-eaters, but will also take insects, spiders, and nectar. They occur in pairs in forest and woodland environments, and their breeding details are similar to the leafbirds.

P. A. CLANCEY

◄ The lesser or Asian fairy-bluebird is a noisy, active, gregarious bird of the treetops that feeds largely on fruit. It is common in lowland evergreen forests throughout Southeast Asia, from India to Java and the Philippines.

D. Avon/Ardea London Ltd

SHRIKES & VANGAS

KEY FACTS

ORDER PASSERIFORMES
SUBORDER OSCINES
FAMILY LANIIDAE
• c. 11 genera • c. 78 species
FAMILY VANGIDAE
• c. 12 genera • c. 14 species

SMALLEST & LARGEST

Coral-billed nuthatch *Hypositta corallirostris*
Total length: 12 cm (4¾ in)
Weight: 10 g (½ oz)

Long-tailed shrike *Corvinella melanoleuca*
Total length: 50 cm (20 in)
Weight: 87 g (3 oz)

CONSERVATION WATCH

!!! The critically endangered species are: Bulo Burti bush shrike *Laniarius liberatus*; São Tomé fiscal shrike *Lanius newtoni*; Uluguru bush shrike *Malaconotus alius*; Mount Kupe bush shrike *Telophorus kupeensis*.
!! There are 4 species listed as endangered: Gabela bush shrike *Laniarius amboimensis*; orange-breasted bush shrike *Laniarius brauni*; Monteiro's bush shrike *Malaconotus monteiri*; Gabela helmet shrike *Prionops gabela*.
! There are 6 vulnerable species.

The wide-ranging, composite family Laniidae (shrikes) is closely related to the small family Vangidae, confined to Madagascar and the Comoro Islands. Recent research indicates that vangas are descended from an ancestral form of helmet shrike from mainland Africa.

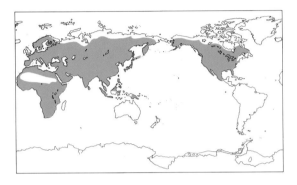

SHRIKES

This family of medium-sized perching birds comprises three distinct groups: the helmet shrikes and the white-headed shrikes (subfamily Prionopinae, nine species), in tropical Africa; the bush shrikes (Malaconotinae, about 44 species), also in tropical Africa, although the black-crowned tchagra *Tchagra senegala* ranges to northwestern Africa; and the "true", predatory shrikes (Laniinae, about 25 species), extensively distributed over much of Africa, Europe, Asia, and North America, but absent from South America, Australasia, and Madagascar.

True shrikes

The true shrikes are patterned in gray, chestnut, black and white, and many of the African species are pied. All have a robust, hooked and notched bill; and except perhaps for the two long-tailed African species (*Corvinella corvina* and *C. melanoleuca*) they are territorial and of solitary disposition. From a vantage perch they watch for terrestrial prey, which is seized on the ground, and many are renowned for their habit of impaling surplus food items on the thorns of bushes or barbed wire fencing. The sexes are well differentiated, and their calls are strident and include snatches from those of other birds. Nests are simple cups of twigs and grasses, lined with rootlets and hair, and these are placed in bushes and trees at no great height from the ground. The eggs are attractively marked.

True shrikes are characteristic of open habitats, preferring steppe—even when quite desertic—and lightly-treed savanna, and virtually all of the birds that breed in the far north (*Lanius* species) migrate to southern latitudes to avoid the winter.

Bush shrikes

The bush shrikes are generally sedentary, mostly in high evergreen forest or savanna woodland in Africa. These birds glean insects from tree limbs and foliage, but some species such as the bokmakierie *Telophorus zeylonus* of southern Africa forage extensively on the ground. Most are attractively colored on the underside, especially the tree-dwelling *Malaconotus* and *Chlorophoneus* species which are largely green on the back, wings and tail, but yellow, orange or even red below—this variation in abdomen color among the *Chlorophoneus* species is now understood to be mimicry of the color patterns of the larger, more powerful and feared *Malaconotus* bush shrikes alongside which they occur, thus gaining advantage for foraging groups of the smaller birds. The nests of all members of this subfamily are constructed in trees or among creepers, and eggs are spotted.

► *The schach or rufous-backed shrike—one of the true shrikes—occurs across southern Asia from Iran to New Guinea. Closely related species inhabit North America, Africa, and Eurasia. Solitary, aggressive birds of open country, shrikes are usually encountered on high conspicuous perches such as telephone poles or isolated dead trees.*

Peter Johnson/NHPA

◄ A white helmet shrike incubating. Helmet shrikes differ from other shrikes in a number of ways, including their gregarious behavior and the rather deep, narrow, and tightly-woven structure of their nests. Exclusively African and totalling about 9 species, helmet shrikes are noisy, conspicuous birds that mainly inhabit arid scrublands.

Helmet shrikes

Helmet shrikes (genus *Prionops*) are characterized by feathering on the crown which forms a brush-fronted "helmet". Moderately gregarious, they roam through the woodland in parties feeding in the trees. The white-headed forms (genus *Eurocephalus*) are also gregarious when on the move or at roost, but they feed independently, behaving much like the true shrikes. Both types construct beautifully fashioned and compacted nests, and among the helmet shrikes some uncommitted adults have been observed helping to rear the young of others of their family group.

VANGAS

The 14 species of vangas are all small, tree-dwelling perching birds found only on islands in the western Indian Ocean. The attractive blue vanga *Cyanolanius madagascarinus,* whose range extends to the Comoro Islands, is the only species to be found outside the main island of Madagascar. A species of doubtful attribution, the kinkimavo *Tylas eduouardi* of Madagascar, long considered to be an aberrant bulbul, is now seen as a member of the Vangidae.

The family exhibits a wealth of bill forms. The sickle-billed vanga *Falculea palliata* of the south-east has a long, scimitar-shaped bill; the three species of the genus *Xenopirostris* have a lower

O. Langrand/Bruce Coleman Ltd

◄ Confined to the Madagascan region, the 14 species of vangas have radiated into a variety of different habitats and ways of life in a manner reminiscent of the finches of the Galapagos or the Hawaiian honeycreepers. This is the rufous vanga.

mandible that is steeply upswept at the end; and the helmet bird *Euryceros prevostii* has a heavy bill which extends far back onto the fore-crown.

While generally viewed as closely related to the Laniidae, vangas differ widely from the true shrikes in their field behavior, and their feeding strategies most closely resemble those of the helmet shrikes. They prefer forest and woodland canopy, gleaning insect and lower vertebrate prey from trees. Many are gregarious, others relatively solitary. Nests are constructed in trees and among creepers, and the eggs are marked with dark spots, but little is known about their breeding activity.

P. A. CLANCEY

WAXWINGS & THEIR ALLIES

Waxwings and their allies have been reclassified into 4 separate families: the waxwings (family Bombycillidae), the silky flycatchers (Ptilogonatidae), the hypocolius (Hypocoliidae), and the palm chat (Dulidae). They are fruit-eating birds of the Northern Hemisphere, and all except the palm chat have silky plumage.

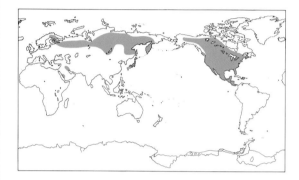

WAXWINGS

The name "waxwing" refers to a red waxlike droplet that forms at the tip of each secondary wing feather, a decoration with no known use. There are three similar species: the Bohemian waxwing *Bombycilla garrulus* in North America, northern Europe and Siberia; the cedar waxwing *B. cedrorum* in North America; and a small population of the Japanese waxwing, *B. japonica,* limited to north-east Asia. Waxwings are fawn-colored birds and are so sleek and trim, with a high crest, that they look carved from wood. Except when nesting they are highly gregarious, and when perched on a limb or wire stand almost touching.

Waxwings prefer northern cedar or evergreen forests for breeding, and their nest is an open cup usually placed high in a tree. The female is fed by the male while she is incubating. Upon hatching the nestlings are fed almost exclusively on insects

J. Kenning

until they fledge at 14 to 16 days. All three species feed heavily on cedar berries, and after the pulp is digested the seeds are passed, making this habit a very important element in reforestation. When nesting is completed waxwings gather into large flocks and move south erratically.

SILKY FLYCATCHERS

This is a New World family, almost restricted to Mexico and southwestern USA. Four species are recognized (in 3 genera), the best-known being the phainopepla *Phainopepla nitens,* a shiny black bird with white wing-patches and crimson eyes, which differs markedly from waxwings in habits, flight and actions. It migrates from dry areas in Mexico to the USA as far north as Sacramento, to nest as single pairs. The nest is built by the male high in a deciduous or evergreen tree, and he also does most of the incubation; both parents feed the nestlings. The pair may seek another territory in a different habitat for their second brood. The period of residency in the USA is very short, from February or March into July—although some birds may overwinter there—then they move as far south as Panama during fall migration. Their most recognizable call is a plaintive high-pitched note given when disturbed by predators or humans.

HYPOCOLIUS

Hypocolius ampelinus, a pale gray bird with black-tipped tail feathers, is limited to the Tigris–Euphrates valley and surrounding areas of Asia Minor. It travels about the scrub country in small flocks feeding almost entirely on small fruits. Male and female build an open cup nest in a pine tree or shrub.

PALM CHAT

The palm chat *Dulus dominicus,* of Haiti and the Dominican Republic in the Caribbean, is a gregarious species which builds communal nests high in palm trees. The structure is up to 1 meter (3¼ feet) in diameter, loosely woven to protect five or more compact individual nests, each with a tunnel to the outside. Incubation and the nestling period are about two weeks, and the noisy birds use the nest structure when resting and roosting.

H. ELLIOTT MCCLURE

▶ *Cedar waxwings. In all three waxwing species, males resemble females in plumage, but juveniles are dull brown and heavily streaked.*

MOCKINGBIRDS & ACCENTORS

Mockingbirds and their allies are thrush-like birds of the New World. Accentors are smaller, similar to sparrows, but with a more slender bill; they live in the Old World, from Europe to Japan.

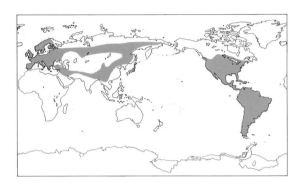

MOCKINGBIRDS AND THEIR ALLIES

The family Mimidae includes 32 species of mockingbirds, catbirds, mocking-thrush, trembler, and thrashers, about equally divided between North and South America. They are sturdy-legged terrestrial or low-vegetation birds, with strong downcurved beaks, short wings, and long tails. Only the blue-and-white mockingbird *Melanotis* *hypoleucos* of South America and the black catbird *Melanoptila glabirostris* of Mexico vary from the usual gray or brown coloration of the family, but many have lovely patterns to distinguish them.

The family is noted for its beautiful singers and mimics. It is tempting to refer to the northern mockingbird *Mimus polyglottus*—which may imitate as many as 20 local bird species—as the most talented, but this may be simply a reflection of the fact that it is an urban species, studied and admired by many observers, whereas those species in less populated areas are not as well known.

All are strongly territorial, especially the mockingbirds which will aggressively attack transgressors whether mammal or bird, and many maintain territories the year round. The North American species live in forest edge, open country, and desert habitats. The successful urbanization of the northern mockingbird, gray catbird *Dumatella carolinensis,* and brown thrasher *Toxostoma rufum* is

KEY FACTS

ORDER PASSERIFORMES
SUBORDER OSCINES
FAMILY MIMIDAE
• 12 genera • 32 species
FAMILY PRUNELLIDAE
• 1 genus • 12 species

SIZE

Mockingbirds (Mimidae)
Total length: 20–33 cm (7¾–13 in)
Weight: 36–80 g (1¼–3⅕ oz)

Accentors (Prunellidae)
Total length: 14–18 cm (5½–7 in)
Weight: 15–43 g (⅗–1⁷⁄₁₀ oz)

CONSERVATION WATCH
!! The Socorro mockingbird *Mimodes graysoni*, Floreana mockingbird *Nesomimus trifasciatus,* and white-breasted thrasher *Ramphocinclus brachyurus* are endangered.
■ The Yemen accentor *Prunella fagani* is listed as near threatened.

Joe McDonald/Tom Stack & Associates

◄Named for the skill with which it mimics the calls of other birds, the northern mockingbird is a common and hardy suburban bird over much of North America. It is strongly territorial, and may lay claim to an ornamental berry bush or similar food resource through the winter, defending it from all comers. An active and fearless bird, it often teases dogs and cats.

▶ *Common and widespread, the brown thrasher resembles the northern mockingbird in many respects, but it is much more strongly migratory, very shy and retiring, and seldom mimics other birds. It spends much of its time on the ground.*

Don & Esther Phillips/Tom Stack & Associates

▼ *Few European birds are so common yet so inconspicuous as the dunnock or hedge sparrow. It spends most of its time on the ground, quietly shuffling and creeping along in a highly distinctive manner, persistently twitching its wingtips. It nests close to the ground, often in ivy or heaps of garden refuse; the female builds the nest and incubates alone, but both parents cooperate in rearing the young.*

Stephen Dalton/NHPA

partly the result of open tree and shrub plantings found in urban areas. These plantings simulate the forest edge conditions preferred by these species and provides them with insects, berries, and fruits. Most build open-cupped bulky nests of twigs, lined with grasses and fibers, in low vegetation. Incubation, mainly by the female but with the male's help, is about two weeks, and the young are fledged at about two weeks. Several species breed a second time each year, and the California thrasher *T. redivivum* may nest as late as fall. Most species forage on the ground, taking terrestrial invertebrates, which they find using their downcurved bill to dig in the soil or search under surface debris. They also eat small fruits in season.

The northern species are migratory to varying degrees, some moving only a few kilometers north or south, others going as far as Central America.

Like insular species elsewhere, those of the Caribbean are endangered through the introduction of small predators and the destruction of environments by humans. The white-breasted thrasher *Ramphocinclus brachyurus* of Martinique and Saint Lucia and the trembler *Cinclocerthia ruficauda* (noted for its habit of shivering and trembling) of the smaller islands of the West Indies are both now rarely seen, but two other species, the scaly-breasted thrasher *Margarops fuscus* of the Lesser Antilles and the pearly-eyed thrasher *M. fascatus,* which ranges from the Bahamas south through many islands, are still quite abundant. These island inhabitants are at risk, however, as they lay only two or three eggs.

ACCENTORS

Limited to Eurasia, 12 species of accentors (family Prunellidae, genus *Prunella*) occupy numerous high-latitude and high-altitude habitats, preferring brushy areas. Most move altitudinally with the season, and some are migratory. They are sparrow-sized with a thrush-like bill and round wings with 10 primary feathers. Not strong fliers, they forage on the ground or in low shrubbery, taking insects during warm weather, but seeds and berries in winter, having a crop and gizzard that can handle this harsh food. The nest, placed on the ground in a crevice or among rocks, is neatly woven and the cup insulated by feathers. The three or four greenish blue eggs are incubated by the female for about 15 days, and the young are quickly fledged, sometimes before they can fly. Usually there are two broods. The dunnock or European hedge sparrow *Prunella modularis* occupies lower scrub country and moorlands in Europe.

H. ELLIOTT MCCLURE

DIPPERS & THRUSHES

KEY FACTS

ORDER PASSERIFORMES
SUBORDER OSCINES
FAMILY CINCLIDAE
• 1 genus • 5 species
FAMILY TURDIDAE
• 48 genera • *c.* 330 species

SIZE

Dippers (Cinclidae)
Total length: 17–20 cm (7–8 in)
Weight: 50–84 g (2–3⅖ oz)

Thrushes & allies (Turdidae)
Total length: 11–33 cm (4½–13 in)
Weight: 12–180 g (½–7⅓ oz)

CONSERVATION WATCH
!!! The critically endangered
species are: Seychelles magpie-
robin *Copsychus sechellarum*;
olomao *Myadestes lanaiensis*;
kamao *Myadestes myadestinus*;
puaiohi *Myadestes palmeri*; taita
thrush *Turdus helleri*; amami
thrush *Zoothera major*.
!! The species listed as
endangered are: black shama
Copsychus cebuensis; Sri Lanka
whistling-thrush *Myiophonus
blighi*; Luzon water-redstart
Rhycornis bicolor; gabela akalat
Sheppardia gabela; Somali thrush
Turdus ludoviciae; spotted ground
thrush *Zoothera guttata*.
! 20 species are listed as
vulnerable.

The five species of dippers making up the family Cinclidae are the only truly aquatic passerine birds. The family Turdidae—thrushes—is an important and widespread family with representatives in virtually every area of the world except the Arctic, Antarctic, and some oceanic islands.

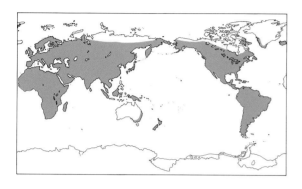

DIPPERS

Dippers (all five species are in the genus *Cinclus*) inhabit clear, swiftly-flowing streams and therefore tend to be found in hilly or mountainous regions throughout much of Europe, Asia, western North America and the northern half of the Andes in South America, wherever such streams occur. Because they do not migrate long distances, however, they do not occupy the streams and rivers of the high Arctic, which might be suitable for them in summer but would freeze over in winter.

Dippers feed largely on the larval forms of aquatic insects such as stone-flies and mayflies, though they take a few small fish as well. They dive into the water, where they either swim or walk along the bottom; when they walk, they usually do so in an upstream direction, holding onto stones with their powerful feet and using the force of the current to keep them on the bottom. They build a bulky, domed nest of mosses near the water, and the female lays and incubates three to six pure white eggs. Both parents raise the young, which take about three weeks to fledge.

Richard T. Mills

◄ *Seldom encountered away from swiftly flowing water, dippers feed underwater on aquatic insects, using their wings to maneuver and bobbing up like corks when surfacing. Even in flight they follow streams, whirring low over the water on rapidly vibrating wings. This is the white-breasted dipper, which is common across Europe and much of temperate Asia.*

THRUSHES

The family Turdidae is commonly divided into two subfamilies: Turdinae, with about 175 species, of which some 64 are classified in the genus *Turdus;* and Saxicolinae with about 155 species.

Subfamily Turdinae

Turdinae includes the typical thrushes, such as the American robin *Turdus migratorius,* the common blackbird *T. merula,* the olive thrush *T. olivaceus* of southern Africa, and the white-necked thrush *T. albicollis* of South America. They are about 23 centimeters (9 inches) long, predominantly quietly colored, mostly in brown with some gray or black, and many are spotted underneath. In many species the sexes are similar in appearance, but in some the males are more brightly colored. Many have fine songs, audible at some distance.

This subfamily has an almost worldwide distribution, although there are few in Australia and none in New Zealand, apart from introduced species. Most live in wooded areas, but they are commonly seen feeding both in trees and on the ground. Most species feed on a wide variety of fruits and animal prey, especially insects and worms; those living in colder climates may vary their diet seasonally, taking fruit in the fall, worms and snails in the winter, and insects during the summer..

The species that inhabit warmer climates are usually resident there throughout the year, whereas those that

▲ The blackbird was once confined to dense forest, but its habitat requirements have relaxed over the past 150 years or so until now it is one of the commonest of birds in European parks and gardens, where it often reaches higher population densities than in its ancestral habitat.

▶ The robin in Britain is a tame, familiar bird of parks and gardens, but it is much more shy and retiring in disposition elsewhere in its range. The red breast is a badge used in territory defense rather than in courtship. Both sexes wear it, and defend territories all the year round—separately in winter, jointly in summer.

breed at higher latitudes migrate quite long distances between their breeding grounds and non-breeding quarters; for example, the redwing *T. iliacus* may breed in northern Scandinavia or Russia and winter in Ireland or Italy. Many of the New World species migrate much further; Swainson's thrush *Catharus ustulatus* breeds in the United States and then flies as far south as northwestern Argentina, while the gray-cheeked thrush *C. minimus*, which breeds largely in northern Canada, migrates as far south as Brazil.

The large majority of species build rather standard, cup-shaped nests, some lining them with mud. They lay two to five eggs, mostly bluish or greenish, and speckled with browns and blacks. They may raise two or more broods in a year. Both parents help to raise the young. The fieldfare *Turdus pilaris*, which breeds in northern Europe, is unusual in that it often nests in colonies. When these large, bold thrushes are threatened by a predator they dive-bomb it, defecating over it as they do so, leaving the would-be predator to withdraw in a sticky mess!

Subfamily Saxicolinae

The members of the other subfamily, the Saxicolinae, are mostly smaller birds—such as the chats, wheatears and robins of the Old World, and the bluebirds of the Americas—averaging 15 centimeters (6 inches) in length. Many are more brightly colored, some having quite vivid reds and oranges, and the bluebirds being predominantly blue. Many of the wheatears are strikingly patterned in blacks, grays, and whites. Males of many species generally have brighter plumage than

their mates. The Himalayan forktails (genus *Enicurus*) have very graduated, deeply-forked tails of black feathers tipped with white. Many species, such as the nightingale *Luscinia megarhynchos*, are fine songsters with a rich range of notes.

This subfamily occurs in a wide range of habitats, from forest and thick scrub (some redstarts, robins, nightingales), at the edges of swift streams (white-capped redstart *Chaimarrornis leucocephalus*), to desert (some of the African chats, some wheatears). They nest in a wider range of sites than the true thrushes, for although some build cup-shaped nests in a bush, many make their nest in a hole in a tree or a cavity under a rock. Many are primarily insect-eaters, others take fruit in season. Many migrate long distances to warmer winter quarters; for example, the wheatear *Oenanthe oenanthe* breeds in the Arctic areas of the Old World, yet all the birds migrate to Africa to avoid the winter. From Greenland they may face a non-stop flight over water of some 3,200 kilometers (2,000 miles).

Also included in this subfamily are several larger species, such as the rock thrushes (genus *Monticola*), and the grandala *Grandala coelicolor,* a bird of the high Himalayas. The male grandala's body is bright blue, whereas the female is dull brown, speckled with white on the head and wings. Grandalas spend most of the year in flocks, sometimes several hundred strong, feeding on the ground. They nest on rocky ledges at an altitude of about 4,000 meters (13,000 feet).

TEXT REVISED BY WALTER E. BOLES

▲ Two forest thrushes of eastern Asia: White's thrush (left), and a white-rumped shama (right). The shama's song rivals that of the nightingale in richness and versatility, but White's thrush is a comparatively quiet and inconspicuous inhabitant of the forest floor.

ORDER PASSERIFORMES
SUBORDER OSCINES
FAMILY TIMALIIDAE
• 54 genera • 288 species
FAMILY TROGLODYTIDAE
• 16 genera • c. 75 species
FAMILY POMATOSTOMIDAE
• 1 genus • 5 species

SIZE

Babblers & allies (Timaliidae)
Total length: 9–36 cm (3½–14 in)
Weight: 10–150 g (²⁄₅–6 oz)

Wrens (Troglodytidae)
Total length: 10–22 cm (4–8⅔ in)
Weight: 8–75 g (³⁄₁₀–3 oz)

Australian babblers
(Pomatostomidae)
Total length: 18–30 cm (7–12 in)
Weight: 35–90 g (1⅖–3⅗ oz)

CONSERVATION WATCH
!!! The species listed as critically
endangered are: gray-crowned
crocias Crocias langbianis; Zapata
wren Ferminia cerverai; Niceforo's
wren Thryothorus nicefori.
!! The endangered species are:
melodious babbler Malacopteron
palawanense; falcated wren-
babbler Ptilocichla falcata; Negros
striped-babbler Stachyris
nigrorum; flame-tempered babbler
Stachyris speciosa; Hinde's pied-
babbler Turdoides hindei;
Apolinar's wren Cistothorus
apolinari.
! 36 species are vulnerable.

BABBLERS & WRENS

The Timaliidae is an Old World family found throughout warmer areas. The Pomatostomidae is an Australasian group that is convergent with, but unrelated to, the true babblers. The wrens, Troglodytidae, are a New World family, except for the winter wren *Troglodytes troglodytes,* whose ancestors presumably crossed the Bering Strait and settled in most of Asia and Europe, where it is known simply as the wren.

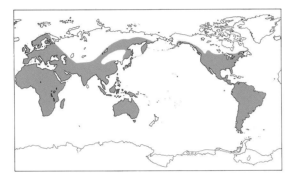

BABBLERS
The Timaliidae comprises three subfamilies: the babblers, Timaliinae, with about 255 species in 49 genera; the parrotbills, Paradoxornithinae, with 19 species in 3 genera; and the bald crows, Picathartinae, with just two species in a single genus.

Babblers
Babblers are widespread throughout the warmer areas of the Old World. In parts of India and Southeast Asia, babblers form a very important part of the bird communities; for example, about one fifth of all bird species in Nepal are babblers. Only one related bird occurs in America: the wrentit *Chamaea fasciata,* which is sedentary on the west coast from Baja California to the Columbia River, Oregon's northern boundary.

A typical babbler is a rather nondescript brown bird about the size of a small thrush. However, they range in size from about 9 centimeters (3½ inches)

► The white-browed babbler lives in arid Australian scrublands in permanent groups of a dozen or so, feeding mostly on the ground. Each group typically consists of a dominant pair and several of their offspring, which assist in raising subsequent broods of young.

Morten Strange

for the pygmy wren-babbler *Pnoepyga pusilla,* a bird with a very short tail, to 36 centimeters (14 inches) for species of laughing-thrushes. While most are predominantly brown, a few are brightly colored with reds, blues, and yellows. The pied babbler *Turdoides bicolor* of South Africa is pure white with black wings and tail. There is very little difference in plumage between the sexes. Most have multipurpose, thrush-like or warbler-like bills, but those in the genus *Pomatorhinus* have longer, more decurved bills, while the slender-billed scimitar-babbler *Xiphirhynchus superciliaris,* which lives in the mountains of northeast India through Burma, is well named for its long, distinctive, decurved bill in the shape of a scimitar.

Babblers inhabit wooded country, many living in thick scrub where they are difficult to locate except by their noisy calls. A few are more specialized to particular habitats, such as the Iraq babbler *Turdoides altirostris,* which lives mainly in reed-beds and other swampy areas, and the Arabian babbler *T. squamiceps,* one of the larger babblers, which lives in scrub along the edges of wadies in desert regions. The majority of species are insect-eaters, hunting for insects or other small invertebrates among the foliage of trees or on the ground. The larger species also take small vertebrates such as lizards. Many will also take berries, and some feed on nectar when it is available.

They tend to be sedentary, defending their territories year-round. Few have been studied in detail, but the following notes seem typical of many: they live in small parties of up to a dozen birds; they are highly social, and the birds remain together almost all the time; in some species they roost together, sitting tightly packed, shoulder-to-shoulder along a branch. The group jointly defends the territory, which the birds do very noisily with a great variety of calls. They breed communally,

▲ *Usually encountered in loose parties of a dozen or so, the blue-winged minla often joins other species of babblers and other birds in flocks wandering through the canopy in mountain forests of India and Southeast Asia. The bluish flight feathers are distinctive but very difficult to see.*

▶ *A popular cage-bird, the red-billed leiothrix is widely known to aviculturists as the Pekin robin. In the wild it lives in small groups that forage on the ground in dense forests from northern India to China.*

► *Winter wren nests are globular structures of moss, twigs, and grass, stuffed into any available cavity and often well hidden. In the breeding season the male winter wren builds a number of nests within his territory. He may pair with several females; once mated, each female chooses one of his nests and raises her brood of young in it, largely unaided by the male.*

the dominant pair building a nest of twigs in a tree or dense bush; the remainder of the group help to defend the pair's nest and raise the young. Young males stay within their own group and breed either by inheriting the territory when their father dies or, if the group becomes large enough, by "budding-off" with some of the other younger birds into a new territory taken from a neighboring, small group. The young females disperse to other groups nearby, presumably to avoid inbreeding.

Many species are widespread and common, some making use of human-made habitats, but others are a cause for concern. Admittedly, their habit of living in dense bush may mean that some have been overlooked and are more common than is believed, but nevertheless, several are known only from small areas of forest, and the rapid removal of this habitat means they are almost certainly threatened. At least five species are restricted to small areas in the Philippines and are considered endangered or threatened; for example, the striped babbler *Stachyris grammiceps* is found only on the slopes of mountains on northern Luzon.

Parrotbills

Parrotbills occur in northern India and Southeast Asia, except for the bearded tit or bearded reedling *Panurus biarmicus* which inhabits reed-beds from Central Asia westwards into Europe. They are small, brownish birds, ranging in size from 10 to 28 centimeters (4 to 11 inches). Most have stubby bills, but that of the spot-breasted parrotbill *Paradoxornis guttaticollis* is particularly deep and rather parrot-like. They generally live in thick scrub, many of them in bamboo thickets.

Bald crows

The two species of bald crow, or rockfowl, are extraordinary birds whose relationship to this and other families has long been debated. Reminiscent of extremely long-legged thrushes, they are bald-headed—hence one of their names—the skin being bright yellow in the white-necked baldcrow *Picathartes gymnocephalus,* and bright blue and pink in the gray-necked bald crow *P. oreas.* They inhabit dense forests in West Africa, where they nest in large caves or on deeply shaded cliffs, building a nest of mud on the rock-face. They tend to nest in groups, perhaps because of the specialized nature of their nest sites. The female lays one or two eggs, and both parents bring insects and worms to their young. Although not thought to be in immediate danger, their special habitat requirements make them vulnerable.

AUSTRALIAN BABBLERS

The five species in the genus *Pomatostomus* are found in Australia and New Guinea. They resemble the babblers of the genus *Pomatorhinus* in the shape of their bills and general behavior. Most are gray, white, and rufous-brown, but the rufous babbler *P. isidorei* of New Guinea is uniformly reddish brown. Hall's babbler *P. halli* was discovered as recently in 1963 in *Acacia* scrub in Queensland.

WRENS

The majority of wren species occur in South and Central America. Just nine species breed in North America, and while several of these migrate south for the winter, almost all other species are sedentary.

Most are small or smallish species, the largest being the cactus wren *Campylorhynchus brunneicapillus,* about 22 centimeters (8⅔ inches) long. They have short, rounded wings and are not strong fliers. All are basically grayish or brownish in color, many heavily streaked with black, and some have white eye-stripes or white throats. Many wrens have powerful voices, and some such as the flutist wren *Microcerculus ustulatus* and the musician wren *Cyphorinus aradus* are highly musical. Some sing antiphonally—a couple of birds giving responses, alternately, to each other. All species build domed nests. Although most of the tropical species are thought to be monogamous, several of the North American species and the winter wren are polygamous, the males building a succession of nests in the hope of attracting a new mate to each. The cactus wren lives in small groups in which juveniles of previous broods help their parents to raise the young of the current brood.

Apolinar's wren *Cistothorus apolinari* is restricted to waterside vegetation in a small area of the eastern Andes in Colombia. The Zapata wren *Ferminia cerverai* is found only in the Zapata Swamp on Cuba, and numbers have been greatly reduced by habitat loss, especially burning.

TEXT REVISED BY WALTER E. BOLES

▼ *Aptly named, the cactus wren often builds its nest amid the formidable spines of the chola cactus. Largest of the wrens, it occurs from southern Mexico north to the southwestern USA.*

Jeff Foott/Auscape International

WARBLERS & FLYCATCHERS

ORDER PASSERIIFORMES
SUBORDER OSCINES
FAMILY SYLVIIDAE
• *c.* 66 genera • *c.* 400 species
FAMILY MUSCICAPIDAE
• *c.* 28 genera • *c.* 179 species

SIZE

Warblers (Sylviidae)
Total length: 9–16 cm (3½–6¼ in)
Weight: 4–70 g (⅐–2⅓ oz)

Old World Flycatchers
(Muscicapidae)
Total length: 10–21 cm (4–8¼ in)
Weight: 7–21 g (³⁄₁₀–⅓ oz)

CONSERVATION WATCH
!!! The critically endangered
species are: taita apalis *Apalis
fuscigularis*; Rodrigues warbler
Bebrornis rodericanus; long-tailed
tailorbird *Orthotomus moreauii*;
long-legged thicketbird
Trichocichla rufa; white-throated
jungle flycatcher *Rhinomyias
albigularis*.
!! The species listed as
endangered are: Pulitzer's longbill
Macrosphenus pulitzeri;
Lompobatlang flycatcher *Ficedula
bonthaina*; furtive flycatcher
Ficedula disposita; Palawan
flycatcher *Ficedula platenae*; ashy-
breasted flycatcher *Muscicapa
randi*; white-browed jungle
flycatcher *Rhinomyias insignis*.
! 45 species are listed as
vulnerable.

M ost of the birds in the family Sylviidae are classified as Old World warblers of the subfamily Sylviinae, which comprises about 349 species in 63 genera. The Muscicapidae, a large, wholly Old World family, is quite unrelated to the Tyrannidae, or New World flycatchers (see page 386). It is divided into two subfamilies: the Muscicapinae, which comprises 153 species in 24 genera; and the Platysteirinae, with 26 species in 4 genera.

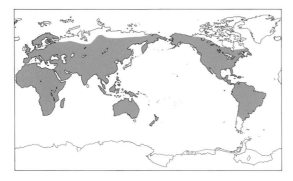

WARBLERS

Very few members of the family Sylviidae have succeeded in penetrating the New World. For example, the Arctic warbler *Phylloscopus borealis* breeds in forests across Siberia and extends just into western Alaska; all the birds migrate to Southeast Asia each winter. Only two species, the golden-crowned kinglet *Regulus satrapa* and the ruby-crowned kinglet *R. calendula*, are widespread in North America.

Within the family Sylviidae is a second, much smaller subfamily, the Polioptilinae or gnatcatchers, containing 12 species in 3 genera. These are restricted to the Americas and do not occur in the Old World. Gnatcatchers are all very small birds, 10 to 12 centimeters (4 to 4¾ inches) long, including their long tails. Like the other warblers, they are insect-eaters. Most occur in South or Central America, but three reach as far north as the USA—the only species that is widespread there, the blue-gray gnatcatcher *Polioptila caerulea*, migrates south for the winter. One South American species, the long-billed gnatwren *Ramphocaenus melanurus*, is 12.5 centimeters (5 inches) long, but this includes its tail measuring 4.5 centimeters (1¾ inches) and an extraordinarily long bill of about 3 centimeters (1¼ inches).

The subfamily Sylviinae is dominated by six genera: *Acrocephalus* (28 species), which is widespread throughout the Old World, especially in reed-beds; *Sylvia* (20 species), most common in Europe and North Africa; *Phylloscopus* (40 species), which forms an important component of forest birdlife, especially in some of the Himalayan region; *Cisticola* (41 species) and *Prinia* (26 species), prevalent in Africa; and *Apalis*

(20 species), exclusive to Africa. Birds in this subfamily have reached almost all of the Old World, including fairly remote islands in the Pacific. In spite of being so widespread, only eight species occur in Australia, including grassbirds and the spinifexbird and only one, the fernbird *Bowdleria punctata*, in New Zealand. Basically they are birds of woodland or scrub, but some, including several members of the genus *Cisticola* and the spectacled warbler *Sylvia conspicillata*, live in very sparse, low vegetation. Many breed in wooded areas at high latitudes or high altitudes, but migrate to warmer climates for the winter, undertaking very long migrations for such small birds. For example, the willow warbler *Phylloscopus trochilus*, which breeds in eastern Siberia, migrates to Central Africa to avoid the winter, making a round trip of some 25,000 kilometers (15,500 miles).

Most warblers are small birds, less than 15 centimeters (6 inches) in length, including the quite long tails of some species. Exceptions are the grassbird *Sphenoeacus afer* of South Africa and

▶ *The golden-crowned kinglet. Kinglets belong to a small genus of tiny, plump warblers characterized by a patch of vivid red or yellow on the crown. They are widespread in the Northern Hemisphere, and most live in coniferous forests.*

Tom & Pam Gardener

the *Cinclorhamphus* songlarks of Australia, both about 23 centimeters (9 inches) long. The majority are dull in color, mostly green or brown, although some are quite heavily streaked with black. In these the sexes generally look alike. An exception is the genus *Sylvia;* the males of many species are brightly colored, with orange or reddish underparts or black and gray patterning; the females tend to be duller. Some, such as the largely African *Apalis* and *Prinia,* have long, strongly gradated tails. In contrast, the crombecs (genus *Sylvietta*) of Africa have almost no tail at all; they climb about on the trunks and branches of trees in a way similar to the nuthatches (family Sittidae). The main food of almost all species is insects, which is probably why so many of those that breed at high altitude or in northern latitudes migrate to warmer areas for the winter. To match this diet, the bill is small and finely pointed. Many warblers also take small fruits and berries when available, and a few take nectar or tiny seeds.

What they lack in appearance, many warblers more than make up for in song, being fine, strong singers. Many indulge in song-flights, soaring up and

"parachuting" down in a striking fashion. Many of the cisticolas, while perhaps not qualifying as fine singers, have elaborate flight displays; indeed many of these are difficult to see in the field, and the species are most easily separated on the basis of their calls.

Most Old World warblers are monogamous, but in some the males are regularly polygamous. For example, the male Cetti's warbler *Cettia cetti* may have as many as five or more females breeding in his territory. The majority build simple cup-nests, often very neatly woven, in thick vegetation. Some build domes or purse-like nests, and the tailorbirds (genus *Orthotomus*) of India and Southeast Asia are renowned for taking two or more large leaves and stitching them together with small fibers or cobwebs threaded through small perforations made in the edges of the leaves; the nest is then built within the leaves. The normal clutch is two to six eggs, depending on the species. The young are raised by both parents (except in polygamous species) and take about two weeks to reach the flying stage, although in some species the young may scatter from the nest before that.

Because many species occur on small islands, their

▲ *Africa is the home of a bewildering range of species of cisticolas, but several forms also occur in Asia and two extend to Australia. Most inhabit grasslands of various kinds. This is the golden-headed cisticola of India, Southeast Asia, and Australia, where it is often known as the tailorbird for its habit of stitching leaves to its nest, aiding in its concealment.*

▲ *A sedge warbler in full song in a German meadow. Although warblers are so plain in plumage that many species are very difficult to identify, they often have loud, rich, and varied songs. Birdsong may be uttered to announce territory – keeping trespassers away – or to attract females, or both; some species have separate songs for each purpose.*

population sizes are often very small and so the species are extremely vulnerable, especially to habitat destruction. For example, in the western Indian Ocean the Aldabra warbler *Nesillas aldabranus*, which was discovered in 1967, inhabits only a tiny area of Aldabra Island; there may be fewer than ten individuals. Slightly less endangered is the Seychelles warbler *Acrocephalus sechellensis*; in 1967 there were probably only 20 to 30 individuals of this species, all on Cousin Island. The island was purchased by the International Council for Bird Preservation (ICBP), and the bird's habitat increased. The numbers have built up well, and some have been distributed to other islands in the Seychelles.

OLD WORLD FLYCATCHERS

Members of the subfamily Muscicapinae occur almost everywhere except in treeless areas, avoiding the center of deserts, high latitudes, and high altitudes. They are primarily birds of wooded areas, from dense forest to very open woodland, almost anywhere they can find an available perch from which to hawk for insects. The orange-gorgetted flycatcher *Muscicapa strophiata* nests at 4,000 meters (13,000 feet) in the Himalayas, where the forest is at its altitudinal limits, but all depend on woodland or shrubs of some sort.

Species that breed in the northern latitudes migrate south to avoid the cold winter, the spotted flycatcher *Muscicapa striata* from Europe going as far as South Africa. The red-breasted flycatcher *M. parva* is unusual in that those from the western end of the range, in Europe, migrate southeastwards to winter in India and Southeast Asia. Those living in warmer areas of the world mostly remain resident throughout the year.

Flycatchers are small birds, about 10 centimeters (4 inches) long, although some have long tails which make their total length much greater than this. They vary markedly in color. Many are rather dull brown birds, but in others the males are more striking—black and white such as the collared flycatcher *Ficedula collaris*, or blue as in the Hainan blue flycatcher *Cyornis hainana*, or other colors. In the vanga flycatcher *Bias musicus* the male is largely black with a white belly and the female is rich chestnut above, with a black head.

As their name suggests, these birds feed mainly on insects. Many literally catch flies, sitting conspicuously on perches and darting out to snap up passing insects. Others forage more among the foliage and take mainly perched insects or caterpillars. In cold weather, when flying insects are scarce, spotted flycatchers may even bring wood-lice to their young. Most have rather broad, flattened bills for catching their flying prey, but some have finer bills and take many of their prey from the ground.

The flycatchers breed in a variety of sites. Many make simple cup-shaped nests in trees or on ledges on cliffs or buildings, whereas others such as the pied flycatcher *Ficedula hypoleuca* nest in holes in trees. Hole-nesting species lay larger clutches than those of open-nesting species, up to eight eggs as opposed to two to five. The pied flycatcher, which winters in Africa and breeds in Europe, is sometimes bigamous: the male displays to a female when she arrives on the breeding grounds and mates with her, but when the female is incubating eggs he may set up another territory nearby and try to attract a second female. The male usually feeds the young in the nest of the first female, so it is the second one that loses out; she gets no help in the rearing of the young, and frequently some die.

A number of species live on islands in Southeast Asia and are poorly known, but possibly endangered. For example, the white-throated jungle-flycatcher *Rhinomyias albigularis* occurs only on the islands of Negros and Guimaras in the Philippines, and the few remaining patches of forest where it lives are being cleared. Many of the inhabitants of West African forests are threatened by continued heavy logging; among these is the Nimba flycatcher *Melaenornis annamarulae*, an all-black flycatcher from the foothills of Mount Nimba.

The other subfamily in the Muscicapidae, the Platysteirinae, is found only in Africa. The two main genera are the puff-backs, genus *Batis,* and the wattle-eyes, genus *Platysteira.* The latter have striking wattles above the eye; these are often red, but in Blisset's wattle-eye *P. blissetti* of West Africa they are green or blue. The birds are found from thick forest to open woodland. They make open cup-shaped nests placed in a fork among branches or on a larger branch and lay two eggs.

Most members of this subfamily seem not to be endangered, but the banded wattle-eye *P. laticincta* occurs only in the Bamenda highlands of West Cameroon, another area of West African forest that is suffering from habitat destruction.

TEXT REVISED BY WALTER E. BOLES

◀ *The rufous-bellied niltava is common in dense forests from the western Himalayas to Burma and Malaysia. Highland populations migrate to lower elevations in winter. Sparrow-sized and inconspicuous despite its brilliant plumage, it resembles other flycatchers in behavior, hawking insects from low perches in dense cover.*

▼ *The forests of India and Southeast Asia are the home of the ferruginous flycatcher. It is an inconspicuous bird most often encountered in the lower branches of the forest canopy, hawking for insects in brief sallies.*

Morten Strange

195

FAIRY–WRENS & THEIR ALLIES

ORDER PASSERIFORMES
SUBORDER OSCINES
FAMILY MALURIDAE
• 6 genera • 26 species
FAMILY ACANTHIZIDAE
• 11 genera • 65 species
FAMILY EPHTHIANURIDAE
• 2 genera • 5 species

SMALLEST & LARGEST

Weebill *Smicrornis brevirostris*
Total length: 8–9 cm (3½ in)
Weight: 5.5 g (⅕ oz)

Rufous bristlebird *Dasyornis broadbenti*
Total length: 27 cm (10½ in)
Weight: 40 g (1⅖ oz)

CONSERVATION WATCH
!! The western bristlebird
Dasyornis longirostris and Biak
gerygone *Gerygone hypoxantha*
are listed as endangered.
! 8 species are listed as
vulnerable.

Australasia—Australia, New Zealand, New Guinea and nearby islands—has a distinctive bird fauna developed during many millions of years of isolation. Although many species derive their common names from European birds with a similar appearance, like the wrens described in this chapter, they are in fact no relation.

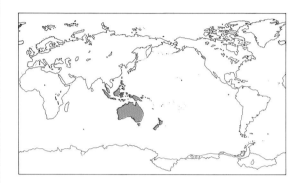

FAIRY-WRENS

The main similarities between the fairy-wrens, grass-wrens, wren-warblers and emu-wrens of Australia and New Guinea (family Maluridae) and the northern wrens is that they are all small and all cock their tails. Otherwise the two groups are quite different. Male fairy-wrens (genus *Malurus*) are among the jewels of the Australian bush, tiny creatures arrayed in stunning combinations of turquoise, red, black, and white. Their mates, and most other Australasian wrens, are more somber. Although the emu-wrens (genus *Stipiturus*), which derive their perverse name from sparse emu-like tail feathers, do have blue bibs, and some of the secretive grass-wrens (genus *Amytornis*) have bold black and chestnut colors, the overall design is for camouflage rather than display.

The family includes 26 species in six genera. All are insectivores, and most forage on the ground or among the underbrush. The wren-warblers of New

► *Alert for danger, a variegated fairy-wren fetches food for its young. As in most fairy-wrens, adult males are clad in glittering blue, females in mousy brown. The variegated fairy-wren is by far the most wide-ranging species, occupying a variety of habitats across Australia.*

Tom & Pam Gardner

Guinea and some fairy-wrens occupy rainforest, where a few species venture into the canopy, but most members of the family are found in grassland or the understory of woodland. The Eyrean grass-wren *Amytornis goyderi,* confined to the sandhills in the driest of Australia's deserts, and several other grass-wrens and emu-wrens use tussocks of spiny grass for protection.

Most members of the family are sedentary and build domed nests in dense vegetation. Like so many Australian birds, the young from one brood frequently remain with their parents to help raise later offspring. Detailed studies of species such as the superb fairy-wren *Malurus cyaneus*, a common bird even in suburban gardens in southeastern Australia, have shown that pairs with helpers are able to rear more young than those without.

AUSTRALASIAN WARBLERS

The Australasian warblers (family Acanthizidae) are a diverse group of inconspicuous small to medium-sized songbirds with slender bills and relatively long legs. The plumage is typically olive, gray or brown, but some thornbills (genus *Acanthiza*) have flashes of yellow or chestnut at the base of the tail, and many gerygones (genus *Gerygone*) have yellow underparts.

Most of the species in this group are found only in Australia, although gerygones, small birds with far-carrying tinkling calls, have spread to New Zealand, many Pacific islands, and much of Southeast Asia. Gerygones are also the only migratory members of the group, most of the others remaining within a territory all year round. Typically the thornbills, whitefaces (genus *Aphelocephala),* and scrubwrens (*Sericornis*) are terrestrial, finding insects and some seeds on the ground or low in the underbrush. A few thornbill species, the weebill *Smicrornis brevirostris* and the gerygones forage in the treetops, while the rock warbler *Origma solitaria* is confined to sandstone outcrops where it builds its domed nest in caves.

Many species breed communally, the younger members of family parties helping to find food for the nestlings and to defend the territory. Studies of marked birds have shown that adults often live more than ten years, with one striated thornbill *Acanthiza lineata* surviving at least 17. Because so many birds survive the winter there is less extra food available in the spring for breeding individuals; this may explain why the Australasian warblers usually lay only one or two eggs in a clutch, whereas in harsher climates similar-sized species often have clutches of a dozen or more.

Most forests contain birds that glean food from the bark of trees. In New Zealand this role is filled by a distinctive group, the Mohouinae (usually classified as a subfamily of Acanthizidae), consisting of the New Zealand creeper *Finschia novaeseelandiae,* the whitehead *Mohioua albicilla,* and the yellowhead *M. ochrocephala.* These are

small birds with spines at the end of their tails. Though they also take food from the leaves, much of their time is spent searching fissures in the bark for insects, sometimes supported by their tails and often hanging upside down. All three form flocks in the winter, and the yellowhead and the whitehead breed communally. The nests of these species are domed; that of the whitehead is invariably inside a tree hollow.

AUSTRALIAN CHATS

Male Australian chats (family Ephthianuridae) are boldly marked in red, orange, yellow, or black and white—colors that stand out in the swamps or arid open shrublands they inhabit—whereas the females tend to be muted versions of their mates. In only the gibberbird *Ashbyia lovensis* are the sexes a similar drab yellow. These five small species have the long legs typical of ground-dwelling birds. Though all the chats have tongues tipped with a brush, which is usually associated with nectar-feeding, their main food appears to be insects. They are sociable birds, often occurring and sometimes nesting in loose flocks. The crimson chat *Epthianura tricolor* and orange chat *E. aurifrons* undertake extensive nomadic movements, sometimes irrupting from central Australia to areas nearer the coast in dry years.

STEPHEN GARNETT

▲ *Two colorful Australians: the orange chat (top) and the splendid fairy-wren. Both are birds of arid interior scrublands, and in both cases females are considerably duller than the males portrayed here.*

▶ *Wedgebills live in loose communities in arid scrublands. They feed on the ground but also spend much time on conspicuous perches, delivering their distinctive calls.*

▼ *The logrunner inhabits rainforests in New Guinea and southeastern Australia. Almost entirely terrestrial, it sifts through leaf litter for insects with wide sideways sweeps of its powerful feet.*

R. Drummond

LOGRUNNERS & THEIR ALLIES

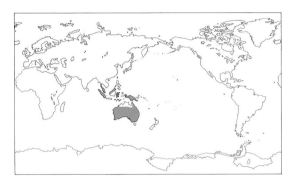

The affinities between the eight genera in the family Orthonychidae are still being examined by scientists, and field guides and other reference books give a variety of common names for them. All but one species live in Australia and New Guinea.

DISTINCTIVE VOICES

In the classification system adopted for this book, the family includes 21 species, generally known as: logrunner and chowchilla (genus *Orthonyx*), whipbirds and wedgebills (*Psophodes*), quail-thrushes (*Cinclosoma*), scrub-robins (*Drymodes*), and in New Guinea the rail-babblers and jewel-babblers (genera *Androphobus, Eupetes, Infrita,* and *Melampitta*). One species occurs in Southeast Asia: the Malaysian rail-babbler *Eupetes macrocerus.*

The logrunner *Orthonyx temminckii* and the chowchilla *O. spaldingii* are birds of the forest floor, solid and medium-sized with stout legs for scratching in the leaf litter while they rest on strong spines extending from the ends of their tail feathers. Although hard to see in their dense habitat, they keep in touch with each other using calls of exceptional volume. Other members of the family have the same habit. The eastern whipbird *Psophodes olivaceus,* a dark olive bird of dense forest, gets its name from the whip-crack call of the male, while the chiming wedgebill *P. occidentalis* and chirruping wedgebill *P. cristatus,* identical but for their song, have ringing calls that carry a long way across the arid shrubby plains they inhabit. Typically these calls are ventriloqual. For instance, despite the strength of scrub-robin calls the birds are very difficult for a predator, or a birder, to find. Most members of the family are colored cryptically, although the complex combinations of black, chestnut, and white in the quail-thrushes of southern and inland Australia are very striking. The only colorful species are the *Eupetes* jewel-babblers of the New Guinean rainforests, the males being a vivid blue, white, and chestnut.

Nearly all species obtain their food on or near the ground. Most take insects, but the chowchilla, at least while nesting, specializes in leeches, and may carry ten or more of these slippery prey at a time. Different species occur in the full range of wooded habitats from Australia to Southeast Asia but reach their greatest diversity in the rainforests where the leaf litter is deep and rich. All appear to be sedentary, with family parties defending the same territories year round. Most species appear to be monogamous; however, at least some, such as the logrunners, breed communally and several non-breeding individuals help to rear the brood.

STEPHEN GARNETT

MONARCHS & THEIR ALLIES

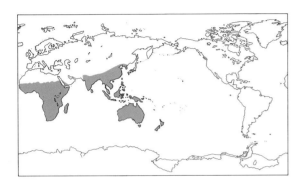

Although the four families of flycatchers and insect-eating birds in this chapter are more diverse in Australia, they also occur in New Guinea, New Zealand, and other islands in the Pacific, with a few farther afield.

FANTAILS

Casual observation would suggest that the gray fantail *Rhipidura fuliginosa* spends more energy catching food than it actually gains from eating it. Its long tail is constantly being fanned or waved from side to side, and the little bird itself is forever pirouetting through the air in pursuit of invisible prey. Yet such behavior must be successful, because all 40 species of fantail (genus *Rhipidura*) from India to New Zealand are a similar shape and size, and most appear hyperactive. The yellow-

bellied fantail *R. hypoxantha* of India is sometimes placed in a separate genus, but the remaining fantails are all combinations of black, gray, white, or rufous, with a tail longer than their body and a short broad bill surrounded by spines called rictal bristles which are thought to increase the catching area of their mouth.

Most fantails forage in or from the foliage at all levels in the woodlands or forest they inhabit. Some are found only in mangroves. In rainforests there is a great diversity, because several species can coexist by feeding at different levels—the canopy, the middle stories, and the ground level. The Australian willie wagtail *R. leucophrys* often feeds on the ground out in the open, for which its long legs are an adaptation.

Those that live in colder parts of Australia or in the Himalayas migrate to warmer climates in the winter, the rufous fantail *R. rufifrons* from southern Australia traveling as far north as New Guinea. Fantails nest in the spring, vigorously defending territories and often rearing several broods in delicate cup-nests. The willie wagtail, one of the

Tom & Pam Gardner

◄ A rufous fantail incubating in an Australian rainforest. Fantail nests are typically slung in low thin horizontal forks, sheltered by foliage.

199

▲ Three representatives of a cluster of related families that reaches its greatest diversity in Australia and New Guinea: the red-capped robin (left) inhabits dry scrublands; the spectacled monarch (center) is a bird of tropical rainforests; and the shrike-tit (right) inhabits eucalypt woodlands.

▼ The willie wagtail has markedly longer legs than other fantails, reflecting the fact that it is the only species to inhabit open country and feed largely on the ground.

Graeme Chapman

better-studied species, threatens rivals by displaying white eyebrows which in defeated contestants are reduced to a mere sliver.

MONARCH FLYCATCHERS

The family Monarchidae comprises more than 90 species in some 17 genera. They are small songbirds with a predilection for flicking their tails and occasionally raising incipient crests, which give many of the species a characteristically steep forehead. Most have long tails, though the Hawaiian elepaio *Chasiempis sandwichensis* has a cocky little tail that it generally keeps erect. Rufous, black, white, blue, or gray, often with iridescent highlights, are the most common plumage colors, but spectacular exceptions include the black and yellow *Machaerirhynchus* boatbills and the paradise flycatchers (genus *Tersiphone*) in which the males have very striking, long tails which may exceed 30 centimeters (11¾ inches) in length. Often the males are more brightly colored than the females, but in many species the sexes are similar.

Most members of the family are forest birds, but there are representatives present in most wooded habitats in Africa south of the Sahara, Southeast Asia, Australia, and islands in the Pacific. The greatest diversity is in New Guinea, where species sally for insects from the very tops of rainforest trees, others feed in the canopy, and many more live in lower parts of the forest. In the frilled monarch *Arses telescophthalmus* there is even a difference in foraging behavior between the sexes, with males searching tree trunks and females sallying after insects in flight. Monarchs often forage in mixed-species parties of birds, which sometimes contain four or five different monarch species moving through the forest together, the insects disturbed by one bird being caught by another.

Most tropical monarchs are probably resident within a territory, but those breeding in temperate regions—such as some populations of the Asiatic paradise flycatcher *Terpsiphone paradisi,* the satin flycatcher *Myiagra cyanoleuca,* and the black-faced monarch *Monarcha melanopsis* of southeastern Australia—migrate to the tropics in the winter. In temperate areas, breeding is restricted to spring, but it is extended in the tropics. Their nests consist of delicate cups sometimes elaborately decorated with lichen.

AUSTRALASIAN ROBINS

The term "robin" has been applied to birds with red breasts all over the world, but the flame, scarlet, rose, and pink robins of Australasia are related to neither their European nor American namesakes. Even more incongruous is the use of the name for other dumpy little flycatchers in the same group; yellow, dusky, and black robins have no red on them at all. Nevertheless they do share many features of the European robin *Erithacus rubecula* in shape, size, and endearing nature.

The Australasian robins are currently classified in the family Petroicidae, comprising 42 species in 11 genera. They are mostly birds of woodland and forest where, rather than search actively, they like to sit and wait for prey to reveal itself. Most catch insects on the ground, although the mangrove robin *Eopsaltria pulverulenta* will take shrimps and small crabs. A few species are more energetic—the lemon-breasted flycatcher *Microeca flavigaster* will glean food from foliage and sally after flying insects, and the Jacky Winter *M. leucophaea* sometimes hovers.

Most members of the family are sedentary, but some of the robins in southern Australia migrate north from Tasmania or out of the mountains during winter. When breeding, the flame robin *Petroica phoenica* and the scarlet robin *P. multicolor* exclude each other from their individual territories; when flame robins migrate, the territories of the sedentary scarlet robins expand. Breeding in temperate areas is in spring, and the nests are usually built with bark, moss, and lichen placed in forks of trees.

The majority of species are common but some are restricted to islands and have become seriously endangered. In New Zealand the Chatham Island robin *Petroica traversi* was reduced to seven birds by 1976 when the population was moved from Little Mangere Island, where the habitat was thought to be highly degraded, to Mangere Island, where the habitat was thought to be better; two of the seven died, but with the help of an innovative fostering program using the closely related Chatham Island tit *P. macrocephala chathamensis*

and transfers to a third island, South East Island, the population has recovered to a point where it no longer needs intensive management.

WHISTLERS, SHRIKE-THRUSHES, AND THEIR ALLIES

Ten genera with 45 species are included in this family (Pachycephalidae), which is centered on New Guinea and Australia. The whistlers, shrike-thrushes, and their relatives are stout flycatchers with sturdy bills, and many have lovely calls. Most are rufous, brown or gray, but among them are some yellow, black and white birds—the shrike-tit *Falcunculus frontatus* and the male golden whistler *Pachycephala pectoralis.* A few, such as the shrike-tit, the crested bellbird *Oreoica gutturalis,* and the crested pitohui *Pitohui cristatus* have crests, but most are unadorned. Most species feed on insects gathered from leaves and branches of the forest or from among the leaf litter, with little of the frenetic activity characteristic of fantails. There is some variation in diet: shrike-tits use their powerful bill to search under bark; the white-breasted whistler *Pachycephala lanioides,* an Australian mangrove species, eats fiddler crabs; and the mottled whistler *Rhagologus leucostigma* of New Guinea eats fruit.

Banding studies suggest that the tropical members of the family are largely sedentary, so pairs occupy a territory year round. The rufous whistler *Pachycephala rufiventris,* however, migrates north after breeding in southeastern Australia, and individuals of the golden whistler certainly disperse over great distances. The golden whistler shows more geographical variation than any other bird species, with more than 70 races described, mostly from isolated populations on Pacific islands.

STEPHEN GARNETT

Hans & Judy Beste/Auscape International

▲ *A frilled monarch on its nest. This monarch inhabits rainforests of New Guinea and Cape York Peninsula, Australia.*

▼ *The white-browed robin is a sedentary species that favors vine scrub and palm thickets along streams in tropical Australia.*

Tom & Pam Gardner

TITS

ORDER PASSERIFORMES
SUBORDER OSCINES
FAMILY AEGITHALIDAE
• 3 genera • 8 species
FAMILY REMIZIDAE
• 4 genera • 13 species
FAMILY PARIDAE
• 3 genera • 58 species

SIZE

Long-tailed tits (Aegithalidae)
Total length: 10–12 cm (4–4¾ in)

Penduline tits (Remizidae)
Total length: 8–11 cm (3½–4¼ in)

Tits, chickadees, and titmice
(Paridae)
Total length: 10–22 cm (4–8⅗ in)

CONSERVATION WATCH
! The white-naped tit *Parus
nuchalis* is listed as vulnerable.
■ 5 species are listed as near
threatened.

▼ *The tail accounts for just over half of
the long-tailed tit's total length of 14
centimeters (5 inches).*

The three families commonly grouped under this heading—the Aegithalidae, the Remizidae, and the Paridae — are widespread, and consist for the most part of small forest-dwelling birds. Some are familiar urban dwellers; the great, blue, and coal tits are among the most intensively studied species in the world.

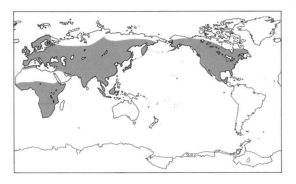

LONG-TAILED TITS

The seven species of long-tailed tits, family Aegithalidae, include the bushtit *Psaltriparus minimus* of western North America and Mexico, and the long-tailed tit *Aegithalos caudatus* which has a very wide range from Britain across Europe and Asia to Japan. The other five species live in the Himalayas and associated mountain ranges. All are tiny, weighing only 4 to 10 grams (¼ ounce) and are 10 to 12 centimeters (4 to 4¾ inches) long, including their long tail. They are mostly dull brownish or grayish, although the long-tailed tit has pink in its plumage and some races have a white head. They have small bills, and feed on small insects and tiny seeds. All species make a beautifully constructed, domed nest, woven from cobwebs and lichens and lined with feathers. Nests are well hidden in thick foliage.

PENDULINE TITS

The penduline tits, family Remizidae, comprise ten species, seven of them (genus *Anthoscopus*) confined to Africa, where they inhabit woodlands. Of the others, the penduline tit *Remiz pendulinus* is widespread across Europe and Asia at low latitudes, where it lives in marshes and along riverbanks; in areas where the winter is particularly severe, it migrates southward. The verdin *Auriparus flaviceps* occurs in the southwestern USA and northern Mexico in dense desert scrub. The tenth species, the fire-capped tit *Cephalopyrus flammiceps*, lives in high-altitude forest in the Himalayas. Most species are largely sedentary. They have small bills and feed on insects and small seeds; the penduline tit takes many seeds from reed mace.

They are mostly rather dull in plumage, although the male penduline tit is quite smartly plumaged, and the male fire-capped tit has, as its name suggests, a bright orange head and throat. They build intricately woven purse-like nests, often suspended from the outer twigs of a bush. The Cape penduline tit *Anthoscopus minutus* is noted for building a nest entrance whose lips can be shut when the bird is in or away from the nest, and to make it even harder for a potential predator, there is a false hole beneath the entrance which is a dead end.

TITS, CHICKADEES, AND TITMICE

The family Paridae comprises 46 species of tits, chickadees, and titmice, all but two in the genus *Parus*. The exceptions are the yellow-browed tit *Sylviparus modestus*, a dull greenish bird with a tiny patch of yellow above the eye, which occurs throughout the Himalayas, and the sultan tit *Melanochlora sultanea*, a black bird 22 centimeters (8½ inches) long, with a striking yellow crest and and yellow underparts which occurs in low-altitude forest from Nepal to the island of Sumatra in Indonesia. Few exceed 13 centimeters (5 inches) in length. They tend to be brown, gray or green above, and paler or yellow underneath. The majority have black caps, some of them with crests, and white cheeks.

Richard T. Mills

The azure tit *Parus cyanus* and blue tit *P. caeruleus* are predominantly blue above.

Species of *Parus* occur throughout most of Europe, Asia, Africa, and North America down into Mexico, but not in the rest of Central America, nor in South America or Australasia. Most species are largely sedentary, although the most northerly populations of species such as the black-capped chickadee *P. atricapillus* in Canada and the great tit *P. major* in northern Europe may move considerable distances to areas with milder winters. They are primarily forest birds. They feed on a wide range of insect and seeds, although all bring insect food to their nestlings; many are seed-eaters through the colder parts of the year. They have short, straight bills; in many species these are slightly stubby and capable of hammering open

small nuts. Those species that live in conifer forests have finer bills, probably associated with their habit of probing into clusters of needles.

All species nest in holes, usually in trees, although some nest in the ground or among piles of rocks. Some species depend on holes left by woodpeckers or natural causes; others excavate their own in rotten wood. Clutches range from three eggs in tropical species to large clutches in temperate species. The blue tit lays the largest clutch of any bird that raises its young in the nest; in oak woodland in central Europe, the average number is 11 eggs, but clutches of 19 have been recorded. Being hole-nesters, some species readily accept nest-boxes and so have been intensively studied.

TEXT REVISED BY WALTER E. BOLES

▲ Wing-feathers spread for braking and feet thrown forward for landing, a blue tit approaches its cavity nest with food for its young. Active, hardy, enterprising, and entertaining, the blue tit is a popular visitor at garden bird-feeders across Europe.

NUTHATCHES & TREECREEPERS

ORDER PASSERIFORMES
SUBORDER OSCINES
FAMILY SITTIDAE
• 1 genus • 25 species
FAMILY NEOSITTIDAE
• 1 genus • 2 species
FAMILY CERTHIIDAE
• 2 genera • 7 species
FAMILY RHABDORNITHIDAE
• 1 genus • 2 species
FAMILY CLIMACTERIDAE
• 2 genera • 7 species

SIZE

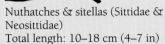

Nuthatches & sitellas (Sittidae & Neosittidae)
Total length: 10–18 cm (4–7 in)

Treecreepers (Certhiidae, Rhabdornithidae, & Climacteridae)
Total length: 12–18 cm (4¾–7 in)

CONSERVATION WATCH
!! The Algerian nuthatch *Sitta ledanti* is listed as endangered.
! The vulnerable species are: beautiful nuthatch *Sitta formosa;* giant nuthatch *Sitta magna;* yellow-billed nuthatch *Sitta solangiae;* white-browed nuthatch *Sitta victoriae;* Yunnan nuthatch *Sitta yunnanensis.*

▼ *A Eurasian treecreeper flies from its nest. Built by both parents, treecreeper nests are usually hidden behind loose bark on a tree-trunk.*

T he families Sittidae (Eurasian nuthatches) and Neosittidae (Australasian sittellas) comprise superficially similar bark-climbing birds. Of the 22 species of nuthatches *(Sitta),* most occur in Eurasia, and there are four species in North America. The wallcreeper *Tichodroma muraria* usually is considered to be an aberrant member of this family. The other three families in this group—the Certhiidae, Climacteridae, and Rhabdornithidae—resemble each other in appearance and behavior, and are called creepers.

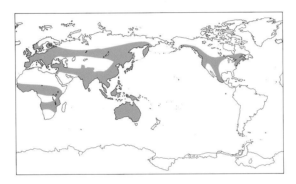

NUTHATCHES AND SITTELLAS

Typical nuthatches are small stocky birds, with longish straight bills. The plumage is typically blue-gray above, and white, pale gray, or reddish brown below; some have a dark cap, and some have a dark streak through the eye area. They are unique in having the ability to run up and down tree trunks, which they do while foraging for insects hidden in the bark. Outside the breeding season many species also take seeds, some specializing on these in winter. Most nest in holes in trees or rocks, often reducing a large entrance to one that they can only just squeeze through, by walling it with mud, dung, and other sticky substances which dry hard.

The two species of sittellas occur in Australia and New Guinea. Although they seem similar to nuthatches in the way in which they clamber about in trees, they are not closely related. They do not nest in holes, but build a beautifully woven cup-shaped nest in the fork of a tree, and camouflage it with pieces of bark attached to the outside. In Australian woodlands, the varied sittella *Daphoenositta chrysoptera* has half a dozen different forms, the colors and patterns varying according to the geographical location; they are now generally recognized as races within the one species.

The wallcreeper inhabits mountain ranges in Europe and Asia, from the Pyrenees to the Himalayas. It nests at high altitudes adjacent to running water, but descends to lower altitudes for the winter. It has bright red in the wings and a long decurved bill with which it feeds on insects.

TREECREEPERS

The most widespread family is the Certhiidae, the Holarctic treecreepers, which includes the tree-creeper *Certhia familiaris* of Eurasia and the brown creeper *C. americana* of North America; five other species are distributed in the Himalayas and associated mountain ranges. These small brownish birds use their stiffened tail feathers for support as they climb up trees. The spotted creeper *Salpornis spilonotus* of Africa and India is sometimes placed in a separate subfamily, Salpornithinae.

The Australian creepers (family Climacteridae) comprise seven species, six of them occurring only in Australia and the seventh in New Guinea. The Philippine creepers (family Rhabdornithidae) comprise only two species, both of which are confined to the Philippines.

TEXT REVISED BY WALTER E. BOLES

Stephen Dalton/NHPA

HONEYEATERS & THEIR ALLIES

KEY FACTS

ORDER PASSERIIFORMES
SUBORDER OSCINES
FAMILY DICAEIDAE
• 6 genera • 55 species
FAMILY NECTARINIIDAE
• 5 genera • 116 species
FAMILY ZOSTEROPIDAE
• 10 genera • 85 species
FAMILY MELIPHAGIDAE
• 42 genera • 167 species

SIZE

Flowerpeckers (Dicaeidae)
Total length: 7–19 cm (2¾–7½ in)

Sunbirds (Nectariniidae)
Total length: 9–21 cm (3½–8¼ in)

White-eyes (Zosteropidae)
Total length: 10–14 cm (4–5½ in)

Honeyeaters (Meliphagidae)
Total length: 7–50 cm (2¾–19¾ in)

F lowerpeckers, sunbirds, white-eyes, and honeyeaters are families of small tree-dwelling birds distributed in Africa, tropical Asia, Australasia, and islands in the southwestern Pacific. The diet of many species includes nectar ("honey") taken from flowers.

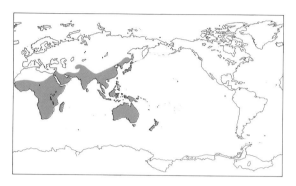

FLOWERPECKERS
The flowerpeckers constitute a family, the Dicaeidae, of about 55 species of birds widespread in southern Asia and Australasia. The family is rather uniform, and at least 40 species are grouped in only two widespread genera (*Prionochilus* and *Dicaeum*). Most flowerpeckers are dumpy little birds with long pointed wings, stubby tails, and short conical bills. The tongue has a characteristic structure, short and deeply cleft, with the edges curled upwards to form twin partial tubes, thought to be an adaptation for gathering nectar from flowers. Many species have dull plumages, but the males of many others have bright colors and patterns, often involving areas of crimson and glossy blue-black. Almost all are strictly arboreal, and feed largely on nectar, fruit, and insects; they inhabit a wide range of environments, from dense rainforest and highland moss forest to arid savanna woodland. Most flowerpeckers build

CONSERVATION WATCH
!!! The 12 critically endangered species are: Cebu flowerpecker *Dicaeum quadricolor*; scarlet-collared flowerpecker *Dicaeum retrocinctum*; white-chested white-eye *Zosterops albogularis*; Mauritius olive white-eye *Zosterops chloronothus*; Kulal white-eye *Zosterops kulalensis*; Seychelles white-eye *Zosterops modestus*; Mount Karthala white-eye *Zosterops mouroniensis*; Rota white-eye *Zosterops rotensis*; Taita white-eye *Zosterops silvanus*; black-eared miner *Manorina melanotis*; Bishop's o'o *Moho bishopi*; Kauai o'o *Moho braccatus*.
!! There are 5 species listed as endangered.
! 27 species are vulnerable.

Morten Strange

◄ *The orange-bellied flowerpecker is widespread in lowland forests of Southeast Asia.*

ge/Flying Colours

Vivacious and intensely active like most sunbirds, the crimson sunbird inhabits India and Southeast Asia. It often hovers to extract nectar directly from small blossoms, but with large flowers such as hibiscus and cannas it usually gains access by piercing the corolla near the base. The call has been likened to a pair of scissors being snapped open and shut.

pendent globular nests of closely matted vegetable fiber, placed in shrubs and saplings.

This family is of particular interest because of the almost symbiotic relationship that exists between certain of its species and the plants (mainly mistletoes) upon which they feed. A good example is the mistletoebird *Dicaeum hirundinaceum* of Australia, which feeds almost entirely on the berries of various species of mistletoes. Its gut is modified to form little more than a simple tube, from which the muscular stomach diverts as a blind sac reached through a sphincter. This arrangement allows berries to bypass the stomach entirely; the fruit pulp is absorbed and digested in passage, but the central seed travels through undamaged, to be voided intact but with a sticky

coating. The seed sticks to the bark on the branches of trees, germinates, and in due course forms a new plant. The relationship thus provides the bird with a reliable source of food, and the plant an effective means of dissemination. So intimate is this partnership that one is seldom found where the other does not also occur.

The common, multicolored pardalotes (genus *Pardalotus*) of Australia closely resemble flower-peckers in size and proportions. They also exhibit an intimate relationship with their food supply: they feed almost entirely on the larvae (and an exudate, known as lerp, produced by these larvae) of leaf-eating insects found on eucalyptus trees. They nest in tree cavities, or in tunnels in the ground. Until recently, the pardalotes, as well as several New Guinean species, such as the crested berrypecker *Paramythia montium,* were included in the Dicaeidae, but DNA studies suggest that pardalotes are part of the thornbill/scrubwren cluster (family Acanthizidae), whereas the New Guinean species may be so aberrant that they belong in separate families (Paramythiidae, Melanocharitidae).

SUNBIRDS

Strongly associated with flowers wherever they occur, sunbirds constitute a family of about 116 species of small, vivacious, colorful birds found through much of the Old World tropics. The group is best represented in Africa, where most habitats support at least one or two species, but one sunbird extends as far east as Australia. Traditionally, the sunbirds have been regarded as a distinct family close to the flowerpeckers (family Dicaeidae) and the honeyeaters (family Meliphagidae), but DNA studies suggest that kinship with the honeyeaters is very distant, while with the flowerpeckers it is so close that the two groups now often are combined in the family Nectariniidae, along with the spiderhunters (genus *Arachnothera*) of Southeast Asia and the sugarbirds (genus *Promerops*) of southern Africa.

Characteristic features include a long, slender, decurved bill with fine serrations along the margins of both mandibles, and a tubular, deeply cleft tongue. A few species are dull, but more typically the male is brilliantly colored, often glittering vivid green or blue above, and red or yellow below; many species have long pointed central tail feathers, and some have brightly colored tufts of feathers on the sides of the breast.

Sunbirds feed on small insects and nectar, and in Africa many flowers are dependent on sunbirds for pollination. The birds sometimes congregate at flowering trees or shrubs, but they are not truly gregarious—males in particular often vigorously defend flowers against all other sunbirds. Their calls are usually abrupt and metallic, but many species have rapid, high-pitched, twittering songs. Sunbirds breed as monogamous pairs and build elaborate pendent domed nests. The male does not help in nest-building or incubation but usually assists in rearing young.

WHITE-EYES

About 85 species of white-eyes occur in Africa and across southern Asia to Australasia, extending north to Japan, and east to Samoa. They occupy almost every kind of wooded habitat, and are often common in suburban parks and gardens. The great majority form a single widespread genus *Zosterops*, whose members are so uniform in appearance and structure that species limits are extremely difficult to determine. Most species are about 10 to 12 centimeters (4 to 4¾ inches) long, unremarkable in proportions with rather slender pointed bills. The plumage is usually green above,

and pale gray or yellow below; there is a distinct ring of tiny, dense, pure-white feathers around each eye. The tongue is brush-tipped, and the birds feed on nectar, fruit, and small insects. White-eyes are also characterized by an almost complete lack of any sexual, age, or seasonal variation in plumage. Traditionally, this family has been viewed as related to the honeyeaters, but studies of DNA hybridization suggest a much closer relationship with the Old World warblers (Sylviidae).

Most are gregarious, living in wandering groups from which pairs periodically drop out in order to breed, returning to the flock when done. These characteristics of sociability and a nomadic tendency confer great powers of dispersal, and white-eyes are successful colonizers of remote oceanic islands. Most islands in the Indian Ocean and western Pacific are occupied by one or another species, and some have two. Sometimes these share the same ancestral form, a result that may occur when the interval between arrivals is sufficiently long to render the original colonists incapable of interbreeding with the new arrivals. Remarkably, two islands in the southwestern Pacific (Norfolk Island, and Lifou Island near New Caledonia) are the site of triple invasions of this kind: on both islands there are three distinct species, all originating from the same ancestral stock.

Graeme Chapman

▲ *The yellow white-eye is a mangrove-inhabiting bird of the coasts of tropical Australia.*

▼ *The regal sunbird (top) is restricted to high mountain ranges in central Africa. The spotted pardalote (bottom) inhabits eucalypt woodlands in southern Australia.*

▲ Two Australian honeyeaters: the eastern spinebill (above) is abundant in thickets of flowering shrubs along the east coast and in Tasmania, and the regent honeyeater (right) is an uncommon and perhaps declining nomad that inhabits dry eucalypt woodlands on the western slopes of the Great Dividing Range.

▼ Coastal heaths of southern Australia are the home of this species, the New Holland honeyeater.

HONEYEATERS

The family Meliphagidae comprises about 167 species of honeyeaters, friarbirds, spinebills, miners, and wattlebirds, all largely restricted to the Australasian region. New Zealand, New Caledonia, and other islands in the southwestern Pacific each have a few species, but most family members inhabit New Guinea (about 63 species) and

Australia (about 68 species). A prominent characteristic of the family is the unique structure of the tongue, which is deeply cleft and delicately fringed at the tip so that it forms four parallel brushes, an adaptation to nectar-feeding. Otherwise honeyeaters vary very greatly in size, structure, and general appearance. All are chiefly arboreal and normally gregarious, but there are very few features common to all.

Some species inhabit rainforest, and these tend to feed mainly on fruit. Many others rely primarily on nectar; and one large genus (*Myzomela,* with 24 species) shows some striking similarities to sunbirds (family Nectariniidae), being very small, vivacious, pugnacious, brightly colored (in the males), and strongly associated with flowering trees and shrubs. But many others rely heavily on insects, and these tend to vary more widely in habitat, general appearance, and behavior.

The honeyeater family forms one of the most prominent elements of the Australian songbird fauna, and it is there that the group reaches its greatest diversity, inhabiting virtually all habitats from highland rainforest to coastal heath and arid scrub in the interior. A birdwatcher taking a casual stroll at any time in any Australian habitat can hardly fail to see many individuals of several species of honeyeaters. Some, such as the New Holland honeyeater *Phylidonyris novaehollandiae,* its close relatives, and the spinebills (genus *Acanthorhynchus*), inhabit coastal heaths and flowering shrubs, where they are colorful and conspicuous in appearance and behavior. Insect-eating forms in woodland and forest are often duller in plumage and less conspicuous in behavior but no less numerous; some glean insects from foliage, while others probe the bark of trees for food. Some species are very specialized in diet and habitat, whereas others are generalists, feeding almost indiscriminately on fruit, nectar and insects in a range of habitats.

Honeyeaters are also notable in the range of social behavior they exhibit. Some are gregarious only in the limited sense that they tend to gather wherever food is abundant, but many others live in permanent structured communities. In the miners (genus *Manorina),* this communalism reaches great sophistication. The situation has been best studied in the noisy miner *Manorina melanocephala,* which usually lives in loose communities of several hundred individuals occupying permanent territories of many hectares; the communal territory is vigorously defended by all members against all avian trespassers (itself a feature virtually unique among birds), but each individual is at the same time a member of a smaller subgroup within the larger community. These subgroups are themselves distinct communities, each consisting of an adult female with several male consorts, which constitute the basic breeding units within the community.

TERENCE LINDSEY

VIREOS

Vireos belong to a strictly New World family, Vireonidae, whose relationships to other songbirds are uncertain. The family was once considered to belong to the large group of "New World nine-primaried oscines"—songbirds in which the tenth (outermost) primary feather of the wing is reduced or absent—but recent research places the vireos well outside this assemblage. Indeed DNA studies suggest a relationship to the Corvidae (crows and jays) and a mostly Old World group of families.

KEY FACTS

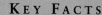

ORDER PASSERIFORMES
SUBORDER OSCINES
FAMILY VIREONIDAE
• 4 genera • 43 species

SIZE

Total length: 10–16 cm (4–6¼ in)
Weight: 8–40 g (³⁄₁₀–1½ oz)

CONSERVATION WATCH
!!! The San Andrés vireo *Vireo caribaeus* is critically endangered.
!! The black-capped vireo *Vireo atricapillus* is endangered.
! The Noronha vireo *Vireo gracilirostris* and Chocó vireo *Vireo* sp. are listed as vulnerable.

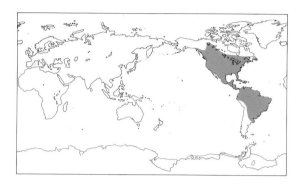

SMALL GRAY-GREEN SONGBIRDS
The 28 typical vireos (genus *Vireo*) are widely distributed in North, Central, and South America. They are small, and most are inconspicuously colored in combinations of dull green, brown, yellow, gray, and white. Of the 14 species that breed as far north as the USA, all but one (Hutton's vireo *Vireo huttoni*, of the western states south to Mexico and Guatemala) are strongly migratory. The most widely distributed species, the red-eyed vireo *Vireo olivaceus*, avoids winter by migrating as far south as Amazonian Brazil, where it shares the equatorial forests with resident vireos. Several months later, others from southern South America migrate northward to the same equatorial forests in the Southern Hemisphere's autumn.

The smallest members of the family are the greenlets (13 species, in the genus *Hylophilus*), which live in forest or scrub, only in tropical regions of Central and South America. Peppershrikes (two species, in the genus *Cyclarhis*) and shrike-vireos (three species, in the genus *Vireolanius*) are quite distinctive, and are classified in their own subfamilies. These are the largest vireos, and are found from Mexico south through tropical South America. The peppershrikes have heavy shrike-like bills. The bills of the shrike-vireos are not as heavy, but this subfamily includes the most brightly colored members of the family.

Vireos occur in a variety of habitats. Some prefer the forest canopy, while others inhabit dense undergrowth, forest edges, or mangroves. They are compulsive singers, often heard during the heat of the day when most songbirds are silent. Vireos are primarily insect-eaters, but many also eat fruits. Their nests, whether high in trees or in low shrubs, are almost always woven cups, suspended by their rims from forked branches. Their eggs are white with small spots of various colors.

KENNETH C. PARKES

John S. Dunning/Ardea Photographics

◀ Striking in appearance but inconspicuous in behavior, the slaty-capped shrike-vireo inhabits tropical lowland forests of South America, favoring the vicinity of streams. It occupies the middle and upper levels of foliage, and frequently joins wandering mixed-species foraging parties of other birds.

BUNTINGS & TANAGERS

KEY FACTS

ORDER PASSERIFORMES
SUBORDER OSCINES
FAMILY EMBERIZIDAE
• 134 genera • 560 species

SMALLEST & LARGEST

Ruddy-breasted seedeater
Sporophila minuta (one of the
smallest species)
Total length: 9 cm (3½ in)
Weight: 8 g (⅓ oz)

White-capped tanager *Sericossypha
albocristata*
Total length: 24 cm (9½ in)
Weight: 114 g (4 oz)

CONSERVATION WATCH

!!! The 6 critically endangered
species are: pale-headed brush-
finch *Atlapetes pallidiceps*;
Venezuelan flowerpiercer *Diglossa
venezuelensis*; Guadalupe junco
Junco insularis; cherry-throated
tanager *Nemosia rourei*; Tumaco
seedeater *Sporophila insulata*;
Entre Riós seedeater *Sporophila
zelichi*.
!! 18 species are listed as
endangered.
! 27 species are listed as
vulnerable.

The family Emberizidae comprises five subfamilies, which formerly were treated as separate families: Emberizinae, Catamblyrhynchinae, Cardinalinae, Thraupinae, and Tersininae. These names are used in the text below because the common names, "buntings", "sparrows", and "finches", apply to birds in different groups in the New World and Old World. All members of the family have wings with nine primary feathers, although a vestigial tenth primary may be present.

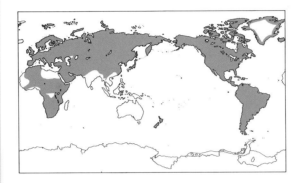

EMBERIZINAE

The Emberizinae includes 279 species in 65 genera, which are thought to have originated in the New World and then dispersed to the Old World in several separate colonizations. The subfamily includes the "sparrows" and various "finches" of the New World, and the "buntings" of the Old World. These birds are small, averaging about 15 centimeters (6 inches) in length, and all have short, conical, attenuated finch-shaped bills reflecting

► *A male yellowhammer at its nest. The female alone incubates the eggs, but both parents cooperate in rearing the young. A typical Old World bunting, the yellowhammer is common in open, scrubby habitats across Europe and western Asia. Strongly gregarious when not breeding, it often joins flocks of other buntings and finches in winter.*

Brian Bevan/Ardea London Ltd

Wardene Weisser/Ardea Photographics

◄ *The rufous-sided towhee varies markedly in appearance across its vast North American range. The eye color varies from red in the north to white in Florida; western birds like this one are conspicuously spotted white on the back, while eastern birds have plain black backs. Towhees inhabit thickets, dense undergrowth, and forest, feeding unobtrusively on the ground.*

▲ *The white-crowned sparrow inhabits grassy clearings and open woodland across northern Canada and alpine meadows in the Rocky Mountains.*

their seed-eating habits. They are mostly terrestrial but may occupy a variety of habitats, including grassland, brushy areas, forest edge, and marsh.

In the Old World there are 42 species, about 28 of which breed in Europe. These include the typical buntings of the genus *Emberiza* which are Eurasian and Afrotropical in distribution, and the Lapland longspur *Calcarius lapponicus* and snow bunting *Plectrophenax nivalis* which breed in Arctic America and Siberia. The latter species breeds in northern Greenland, farther north than any other land bird. Most species tend to have brown streaked body plumage. The pointed crest and glossy blue-black plumage of the male crested bunting *Melophus lathami* and the dark-blue plumage of the Chinese blue bunting *Latoucheornis siemsseni* are unusual in the group.

In the New World there are 234 species in about 60 genera. The sparrows (especially in the genera *Zonotrichia* and *Ammodramus*) are mostly brown and streaked. But many species are very attractively colored—for example, the rufous and green of some towhees (genus *Pipilo*) and brush-finches (genus *Atlapetes*), and the bright orange-yellow of the saffron finches (genus *Sicalis*) of tropical Central and South America.

Emberizines are generally seed-eaters but often switch to a diet of insects when feeding their young. In a few species of juncos (genus *Junco*) and crowned-sparrows (genus *Zonotrichia*) of the New World, bill lengths have been shown to change

with the seasons, being shorter when feeding on seed. Some of the New World genera (for example, *Melospiza, Pipilo, Zonotrichia*) feed on the ground with a "double kick"—they jump back with both feet at once, raking the soil to reveal a tasty morsel. None of their Old World cousins feeds in this manner. Birds in the grasslands of tropical Central and South America (for example *Sporophila, Tiaris, Geospiza*) may clamp food items onto a perch with their feet and then pull off pieces with their bill.

Most emberizines build a cup-shaped nest of grasses, roots and other plant fibers. However, members of several genera in tropical America (*Tiaris, Loxigilla, Geospiza, Melanospiza*) build domed nests; and the Cuban grassquit *Tiaris canora* often constructs a long tubular entrance to its nest. Most species tend to be monogamous or occasionally bigamous. Biochemical studies of blood proteins have revealed that in the white-crowned sparrow *Zonotrichia leucophrys*, which was thought to be monogamous, females engage in "extramarital affairs" so that some 40 percent of their offspring are "illegitimate".

With a few exceptions—for example, the song sparrow *Melospiza melodia* and the five-striped sparrow *Aimophila quinquestriata*—most species tend to possess small song repertoires consisting of short simple utterances. A few species are well known for their regional song dialects—for example, the ortolan bunting *Emberiza hortulana* in the Old World, the white-crowned sparrow in

▲ The red-legged honeycreeper (top) is widespread from Mexico to central Brazil. It is a common, vivacious, and gregarious forest bird that feeds on nectar, fruit, and small insects. The superb tanager (right) is confined to the forests of Brazil. The painted bunting (bottom) breeds in thickets and weedy tangles across the southeastern United States from Texas to North Carolina; it winters in Central America. In the superb tanager the sexes are similar, but in the other two species the female is very much duller than the male.

cardinals (genus *Cardinalis*), the buntings (*Passerina*) and some of the grosbeaks (*Pheucticus*); but in the 12 *Saltator* species of Central America, the West Indies, and South America, both sexes are similar in color, being mostly green and brown.

All cardinalines are thick-billed finches, and they differ in structure of palate, tongue and jaw musculature from the more slender-billed emberizines, described earlier. The former are adapted to crush seeds, whereas the latter tend to be seed-peelers, removing the husks and swallowing the kernels whole. The diet of cardinalines consists of a mixture of fruit, grain, and insects; the saltators tend to feed on berries, fruit, and insect larvae.

All species build cup-shaped nests placed in trees or bushes. Males of the black-headed grosbeak *Pheucticus melanocephalus* and rose-breasted grosbeak *P. ludovicianus* assist the female in incubation and have the peculiar behavior of singing while sitting on the eggs. The cardinal *Cardinalis cardinalis* is well known for its regional song dialects.

Some of the species breeding in North America, such as the dickcissal *Spiza americana* and the indigo bunting *Passerina cyanea*, winter in Central America; and the latter has been the subject of fascinating studies on various aspects of migration, especially orientation using the stars and the Earth's magnetic field.

THRAUPINAE
The tanagers, about 240 species in 58 genera, are entirely New World in distribution. Four species (genus *Piranga*) breed in the USA and migrate to the tropics in the fall; some 163 species are confined to South America. The Andes is the center of radiation for the group. Most (about 149 species) are forest-dwellers, whereas others (about 54 species) tend to prefer semi-open areas; the rest do not show obvious habitat preferences.

Although a few species are drab and secretive, the tanagers include some of the most colorful of all birds, especially in the genus *Tangara*. Most are primarily fruit-eaters but they also take insects. Some such as the honeycreepers (genera *Cyanerpes* and *Dacnis*) and flower-piercers (*Diglossa*) are nectar specialists and have long thin delicate bills. A few (for example, the genera *Lanio* and *Habia*) are insect specialists, and have heavy bills equipped with notches to better grasp insects. Some tanagers are noted for their behavior of following troops of army ants that stir up insects in their wake, and others are known to participate in mixed-species flocks. Some species have attractive songs. Upon seeing a potential predator, the thick-billed euphonia *Euphonia laniirostris* imitates the alarm calls of other songbirds, for example the variable seedeater *Sporophila aurita*. This attracts the latter species to come and harrass ("mob") the predator, leaving the euphonia to go about its

North America, the rufous-crowned sparrow *Zonotrichia capensis* of tropical America, and the 14 Galapagos finches (genera *Geospiza, Cactospiza, Platyspiza, Camarhynchus, Certhidea, Pinaroloxias*) made famous by Charles Darwin. The *Melospiza* and *Zonotrichia* sparrows have been used in extensive studies of song-learning, which have helped scientists to understand the relative roles of learning versus inheritance in vocal traditions.

CATAMBLYRHYNCHINAE
The plush-capped finch *Catamblyrhynchus diademata* of the Andes of South America is a finch-like bird about 15 centimeters (6 inches) long, and its stiff erect golden-brown crown feathers give it its name. This species favors bamboo groves, where it forages on insects and vegetable matter; its bill is similar in shape and structure to some of the Old World parrot-billed babblers (genus *Paradoxornis*), which also are bamboo specialists. Little is known of the plush-capped finch's life history. In some classification systems it is placed with the tanagers in the subfamily Thraupinae.

CARDINALINAE
This group consists of 39 species in 9 genera, and is distributed throughout the Americas. Some, notably the "buntings" (genus *Passerina*), are very colorful. The males may have much brighter plumage than the females—for example, the

Ian Beames/Ardea London Ltd

business and avoid a dangerous situation.

Most tanagers build cup-shaped nests placed among moss or dead leaves. Others, such as the *Euphonia* and *Chlorophonia* species, build globular nests. The female does most of the nest-building and incubation. Some males may feed the incubating female, and both sexes feed the young. Incubation is shorter (10 to 13 days) for the open cup-nesters, which are subject to higher predation rates, and longer (18 to 24 days) for the *Euphonia* species, which build domed camouflaged nests. Breeding pairs and their broods form small post-breeding flocks of up to a dozen individuals. Five species are known to have helpers at the nest.

TERSININAE

The swallow tanager *Tersina viridis,* which is merged with the thraupine tanagers in some classification systems, occupies a range from Panama to the northern parts of South America and Trinidad. It is a bird of montane areas and is migratory in at least part of its range. It is about 15 centimeters (6 inches) long, with longer wings and shorter legs than the thraupine tanagers, and is endowed with a wide, flattened bill, slightly hooked at the tip, and a distinct palate. The bill is adapted for digging and for capturing insects on the wing in the manner of swallows (hence its name). It also eats fruit which it plucks from trees; large pulpy fruit is preferred to small berries, which are rarely taken.

Courtship consists of a unique curtsey display

with flattening the head feathers, lowering and quivering of wings, and bowing. This display is also used in aggressive encounters, notably with territorial neighbors. Song is delivered only by the male, mostly during the nest-building phase. A cup-shaped nest is placed in a horizontal burrow excavated in a bank or in a hole in a wall. The swallow tanagers may also take over abandoned burrows. The female alone incubates the three eggs, but the male may assist in feeding the young.

LUIS F. BAPTISTA

▲ *The paradise tanager is a bird of the treetops in South American tropical forests.*

▼ *The Brazilian tanager is confined—as its name suggests—to Brazil, where it inhabits forest edges and clearings.*

John S. Dunning/Ardea London Ltd

KEY FACTS

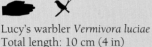

ORDER PASSERIFORMES
SUBORDER OSCINES
FAMILY PARULIDAE
• 22 genera • 114 species
FAMILY ICTERIDAE
• 25 genera • 92 species

SMALLEST & LARGEST

WOOD WARBLERS (PARULIDAE)

Lucy's warbler *Vermivora luciae*
Total length: 10 cm (4 in)
Weight: 6 g (⅕ oz)

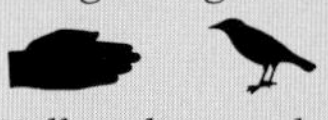

Yellow-breasted chat *Icteria virens*
Total length: 19 cm (7½ in)
Weight: 33 g (1 oz)

ICTERIDS (ICTERIDAE)

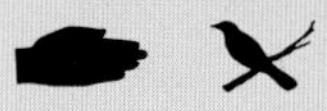

Orchard oriole *Icterus spurius*
Total length: 15 cm (6 in)
Weight: 18 g (⅗ oz)

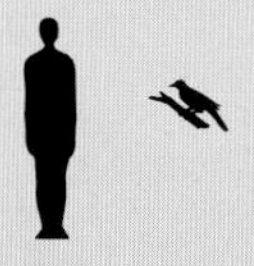

Olive oropendola *Psarocolins bifasciatus*
Total length: 52 cm (20½ in)
Weight: 445 g (15⅗ oz)

CONSERVATION WATCH
!!! The critically endangered species are: gray-headed warbler *Basileuterus griseiceps;* Semper's warbler *Leucopeza semperi;* Paria whitestart *Myioborus pariae;* Bachman's warbler *Vermivora bachmanii;* Forbes's blackbird *Curaeus forbesi.*
!! There are 7 species listed as endangered: golden-cheeked warbler *Dendroica chrysoparia;* saffron-cowled blackbird *Agelaius flavus;* yellow-shouldered blackbird *Agelaius xanthomus;* Baudó oropendola *Gymnostinops cassini;* red-bellied grackle *Hypopyrrhus pyrohypogaster;* Martinique oriole *Icterus bonana;* pampas meadowlark *Sturnella militaris.*
! There are 8 species listed as vulnerable.

WOOD WARBLERS & ICTERIDS

These two families inhabit the Americas. The birds of the wood warbler family, Parulidae, are commonly known as warblers, redstarts, waterthrushes, ovenbirds (quite distinct from the birds of the same name that belong to the family Furnariidae), and yellowthroats. The most morphologically and ecologically diverse group of the New World songbirds is the family Icteridae, for which there is no really good comprehensive English name. Often called "American blackbirds", relatively few of its approximately 92 species are wholly or predominantly black.

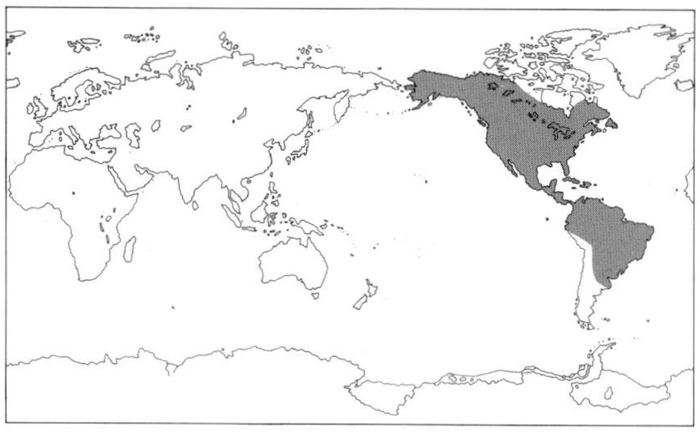

WOOD WARBLERS

Wood warblers reach their greatest diversity in North America, Central America, and the West Indies; about 13 species and many additional subspecies are actually confined to islands in the West Indies. However, members of the family occupy virtually every conceivable habitat south of the tundra. Most nest in trees, shrubs, or vines; those that nest on the ground tend to do so in wooded habitats rather than open country. Several species are typical of the belt of coniferous forest stretching across Canada and the northern USA, extending south in the mountains to Mexico. Others prefer deciduous and mixed forests. The yellowthroats (genus *Geothlypis*) are unusual in nesting in marshes and upland areas that may lack any woody vegetation. Habitat requirements for some species are quite restrictive: some populations of the most widely distributed species, the yellow warbler *Dendroica petechia,* are confined to coastal mangroves. Kirtland's warbler *Dendroica kirtlandii* breeds only in middle-sized jack-pines in Michigan, USA. The Louisiana waterthrush *Seiurus motacilla* of temperate eastern USA seldom nests far from running water.

Although primarily insect-eaters, wood warblers exhibit many foraging methods. Some are excellent flycatchers and have evolved the flatter bills surrounded by the longer hair-like rictal bristles typical of birds that forage aerially. Narrow, thin bills exemplify the many species that pick small insects or their eggs from leaves or twigs. One species, the black-and-white warbler *Mniotilta varia* of eastern North America, has creeper-like habits—climbing on trunks and limbs of trees to search crevices in the bark for small insects and eggs—and has evolved a bill, toes, and claws longer than those of its nearest relatives.

Most of the wood warblers are brightly colored and patterned. Among the North American species, males in the breeding season are often more brilliantly colored than females, but both sexes usually molt into a less-conspicuous plumage before the autumn migration, similar to the plumage of their young; in fall, many conspicuous details of the breeding plumage are lost, so these species present notorious identification problems for birdwatchers and are known as "confusing fall warblers". Among the few dull-plumaged northern species in which the sexes are alike year-round are the waterthrushes and ovenbird (all in the genus *Seiurus*), the worm-eating warbler *Helmitheros vermivora,* and Swainson's warbler *Limnothlypis swainsonii.*

In most of the South American species the sexes are alike in color and do not have obvious seasonal plumage changes. Most of these belong to the large tropical genera *Myioborus* and *Basileuterus,* which barely reach the southwestern USA.

ICTERIDS, OR AMERICAN BLACKBIRDS

The family Icteridae is divided into three distinctive subfamilies. Oropendolas (genus *Psarocolius*), caciques (four genera) and American orioles (genus *Icterus,* including the troupial *I. icterus*) constitute the subfamily Icterinae; most are tree-dwelling, and all (except one cacique) build woven pendulous nests. The subfamily Agelaiinae includes the "blackbird" members of the family as well as meadowlarks (genus *Sturnella*), which are terrestrial birds of grasslands, grackles (genus *Quiscalus*), cowbirds (genera *Molothrus* and *Scaphidura*), and several small genera endemic to South America. Members of this subfamily differ from virtually all other songbirds by molting their tail feathers in a centripetal sequence, that is from the outermost to the innermost. The bobolink *Dolichonyx oryzivorus* is the sole member of the subfamily Dolichonychinae. It breeds in the grainfields and marshlands of Canada and the northern USA, and migrates to Brazil and Argentina. It is the most bunting-like in appearance, and has two complete molts annually (very rare in birds in general).

Although the family is predominantly tropical, it has numerous representatives in both the northern and southern temperate regions. The red-winged blackbird *Agelaius phoeniceus,* one of the most abundant of North American birds, breeds as far north as Alaska—as does the rusty blackbird *Euphagus carolinus.* At the other end of the world, the long-tailed meadowlark *Sturnella loyca* breeds as far south as Tierra del Fuego and the Falkland Islands.

Except for some of the oropendolas and caciques, relatively few members of this family are true forest birds. Many inhabit swamps, marshes, and savannas, and many of the orioles prefer open woodlands and arid scrub. Some grackles have become highly urbanized; the common grackle *Quiscalus quiscula* is abundant in city parks in much of the USA and Canada, and large noisy roosts of the great-tailed grackle *Q. mexicanus* are

typical of towns and villages throughout Mexico and Central America. Size difference between males and females is especially conspicuous in this family, particularly in the larger species—in the great-tailed grackle the male may weigh as much as 60 percent more than the female. In fact, this family has a greater range of sizes among species than any other family in the order Passeriformes.

Nesting habits are highly diverse. Some of the oropendolas, caciques, and marsh-inhabiting blackbirds choose to nest in dense colonies. Only one of the six species of cowbird (genera *Molothrus* and *Scaphidura*) builds its own nest; all of the others are parasitic, depositing eggs in the nests of other birds. Many host-bird species have evolved defenses against this parasitism, but when cowbirds move into an area they haven't previously inhabited, the new host species, especially if already rare, may be severely affected. This happened when the brown-headed cowbird *Molothrus ater* invaded the range of Kirtland's warbler *Dendroica kirtlandii* in Michigan and the black-capped vireo *Vireo atricapillus* in Oklahoma, and when the shiny cowbird *M. bonariensis* spread into the limited area occupied by the yellow-shouldered blackbird *Agelaius xanthomus* in Puerto Rico.

KENNETH C. PARKES

▲ A male bobolink in its summer home; the female is dull, brownish, and streaked. This is a bird of lush grassland, including hay meadows and clover fields, across temperate North America. It winters in South America. Its rich, bubbling song—often uttered in flight—has a distinctive banjo-like timbre.

Townsend P. Dickinson/Comstock

◄ One of the loveliest of wood warblers, the prothonotary warbler is seen to best advantage in its preferred breeding habitat, the swamps and flooded forests of the southeastern United States, where its orange-yellow head and breast glow against the gloom of the cypress trees. It spends most of its time near the ground and, like most wood warblers, it winters in Central and South America.

FINCHES

KEY FACTS

ORDER PASSERIFORMES
SUBORDER OSCINES
FAMILY FRINGILLIDAE
• 20 genera • 122 species
FAMILY DREPANIDIDAE
• 17 genera • 29 species
FAMILY ESTRILDIDAE
• 26 genera • 123 species
FAMILY PLOCEIDAE
• 15 genera • 136 species
FAMILY PASSERIDAE
• 3 genera • 33 species

SMALLEST & LARGEST

Orange-breasted waxbill
Amandava subflava
Total length: 9 cm (3½ in)
Weight: 7 g (¼ oz)

Long-tailed widow *Euplectes
progne*
Total length: 60 cm (24 in)
Weight: 42 g (1⅗ oz)

CONSERVATION WATCH

!!! There are 7 critically
endangered species: nukupuu
Hemignathus lucidus; poo-uli
Melamprosops phaeosoma; Oahu
alauahio *Paroreomyza maculata*;
ou *Psittirostra psittacea*; São Tomé
grosbeak *Neospiza concolor*;
Mauritius fody *Foudia rubra*;
Ibadan malimbe *Malimbus
ibadanensis*.
!! The 14 endangered species
include: akepa *Loxops coccineus*;
Hawaii creeper *Oreomystis mana*;
red siskin *Carduelis cucullata*;
yellow-throated serin *Serinus
flavigula*; Gouldian finch *Chloebia
gouldiae*; pink-billed parrotfinch
Erythrura kleinschmidti; Gola
malimbe *Malimbus ballmanni*.
! There are 31 vulnerable species.

T he word "finch" commonly describes several groups of seed-eating birds, divided into five families. Although typical finches have a cone-shaped bill, not all members of these five families are seed-eaters and indeed some of them have very different bill shapes.

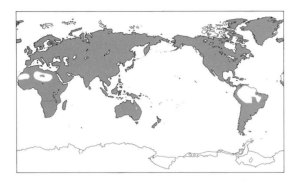

WINGS WITH NINE PRIMARIES

The chaffinches and cardueline finches (family Fringillidae) and the Hawaiian honeycreepers (family Drepanididae) have only nine large primary feathers in each wing (the tenth primary is vestigial) and are therefore grouped with the mainly American assemblage of songbirds called "New World nine-primaried oscines". The geographic name for this assemblage is hardly appropriate, because the center of diversity in the family Fringillidae is outside the New World, and the Hawaiian honeycreepers occur only in the Hawaiian islands. Nonetheless, anatomical and genetic studies indicate that these two families are close relatives of the Old World buntings and of the other members of the nine-primaried assemblage—the strictly New World tanagers, wood warblers, cardinal grosbeaks, and icterine blackbirds and orioles.

Fringillid finches and Hawaiian honeycreepers are each other's closest relatives, a relationship that belies their marked differences in appearance, distribution, and behavior. All fringillid finches have conical bills, and most inhabit temperate climates, are migratory, and eat large quantities of seeds and buds. In contrast, the Hawaiian honeycreepers are restricted in distribution, exhibit an incredible range of bill sizes and shapes, inhabit tropical and semi-tropical forests, and many eat nectar from flowers.

► The crossed mandibles of the crossbill are used to lever seeds from the cones of pine and spruce trees, upon which the birds feed almost exclusively.

Denis Avon/Ardea London Ltd

The only obvious external similarity linking Hawaiian honeycreepers with the cardueline subfamily of the Fringillidae is the possession of red or yellow in the plumage at some stage of the life cycle. However, similarities in muscle patterns, protein structure, and DNA sequences do unite these two apparently disparate families.

FRINGILLID FINCHES

The family Fringillidae is made up of two subfamilies: chaffinches and cardueline finches.

Chaffinches

The chaffinches (subfamily Fringillinae) are the only fringillid finches that lack red or yellow in their plumage. There are three species. The common chaffinch *Fringilla coelebs* and the brambling *F. montifringilla* are primarily Eurasian, inhabiting open woods, gardens, and farmlands; they have boldly patterned males, but the females are duller. The blue chaffinch *F. teydea* is confined to the Canary Islands, where it inhabits mountain pine forests; the males are evenly colored blue. Chaffinches feed on seeds, buds, fruits, and insects.

The common chaffinch has a distinctive place in the history of song research in ornithology. It was the first species in which the interplay between song, innate vocal drives and learning was studied extensively. By playing recordings of songs at specific times during the growth of chaffinches, ornithologists discovered that a learning period early in life lasts roughly a year and is followed by a silent period. After the silent period, the birds engage in a period of practice, during which their vocalizations crystallize into a song. The stages of song development are closely linked to levels of endocrine hormones.

Cardueline finches

The 119 species of cardueline finches (subfamily Carduelinae) are distributed widely in the Americas, Africa, and Eurasia, with the greatest diversity in the Himalayan region. Most species occur in subarctic, temperate, or desert regions; the few that occur near the Equator inhabit mountainous regions with temperate climates. They do not breed natively in Madagascar, the Indian subcontinent south of the Himalayas, or Australasia, although humans have introduced them to many non-native regions and islands such as Bermuda and New Zealand. The genera in this subfamily include *Serinus* (commonly known as canaries and seed-eaters), *Carduelis* (greenfinches, goldfinches, redpoll, siskins), *Acanthis* (redpolls, twite, linnet), *Leucosticte* (rosy finches), *Carpodacus* (rosefinches), *Pinicola* (pine grosbeaks), *Loxia* (crossbills), *Pyrrhula* (bullfinches), *Coccothraustes* (hawfinch, evening grosbeak), and 11 others.

Most Northern Hemisphere cardueline finches are migratory, and many species exhibit nomadic behavior. Certain nomadic species such as the

S.R. Cannings

▲ A male American goldfinch tests the water. Bathing, in either water or dust, is a vital part of routine feather care and maintenance for birds. This goldfinch is one of many that use shallow, secluded pools.

◄ The goldfinch (top left) and the chaffinch (lower left) are two of the commonest and most widely-known finches of European fields and gardens.

crossbills will breed at any time of the year, as long as food is abundant. This nomadic tendency may in part explain the ability of the group to colonize distant oceanic islands—in prehistoric times a cardueline-like colonist gave rise to the range of Hawaiian honeycreepers.

Cardueline finches eat primarily seeds, buds, and fruits. Insects do not make up a substantial proportion of the diet, even during the nesting season. This is distinctive, because most grain- and fruit-eating songbirds switch over to insects during nesting. The carduelines, in contrast, feed their nestlings a mix of predigested plant material, including seeds and buds. They typically nest well off the ground in shrubs and trees, although a few species have developed close commensal relationships with humans (for example, the house finch *Carpodacus mexicanus* of North America) and may nest on buildings. An unusual aspect is that cardueline young defecate in the nest, so that the fecal material builds up on the walls and the rim; in other birds, the young either eject feces out of the nest or package fecal material in a sac that is removed by the adults. Although canaries are the most renowned cardueline singers, many other species produce beautiful songs. Domesticated canaries are descendants of the canary *Serinus canaria,* a species endemic to the Azores, Canary,

and Madeira islands. Brought to Europe as a cage-bird in the sixteenth century, the species has subsequently been bred with other carduelines with pleasant songs, such as the bullfinches; the resulting hybrids often have exceptional singing abilities. Hybrids between different species of carduelines are rare in nature, yet in captivity many hybrids have been produced, including several from genetically distant parental species.

THE HAWAIIAN HONEYCREEPERS
Several Hawaiian honeycreepers are quite similar in appearance to the fringillid finches, probably because of the close evolutionary relationship between the two families. In the family Drepanididae there are 29 extant or recently extinct species in 17 genera.

Finch-like Hawaiian honeycreepers
The finch-like honeycreepers (subfamily Psittirostrinae) consist of nine modern species; of these, only five are extant, and even they survive in but small numbers. Each of the finch-like Hawaiian honeycreepers has a conical bill; in one species, the Maui parrotbill *Pseudonestor xanthophrys,* the bill is modified into a structure remarkably like the bill of a small parrot.

Thirteen more species (in four genera) are sometimes classified in a separate subfamily (Hemignathinae); of these, two species are extinct, two are probably extinct, and the remaining nine persist in very small numbers or in greatly reduced distributions. These birds resemble the finch-like honeycreepers but have less conelike bills, which are either relatively straight, as in the Kauai creeper *Oreomystis bairdi,* or have the upper portion of the bill greatly elongated and curved, as in the common amakihi *Hemignathus virens.*

All finch-like Hawaiian honeycreepers eat seeds, buds, and insects; and two of them, the Laysan finch *Telespyza cantans* and the Nihoa finch *T. ulima,* also include the eggs of seabirds in their diets. The calls and songs are poorly known or inadequately studied, but are variously described as canary-like or linnet-like. Of the species for which there is nesting information, nests are formed of grass, stems, and rootlets; the Laysan finch places it up to 10 centimeters (4 inches) above the ground in grass tussocks, the Nihoa finch in rocky outcroppings of cliffs or crevasses, and the palila *Loxioides bailleui* in trees. The breeding season of the honeycreepers on the leeward islands—the Laysan and Nihoa finches—begins about January and appears to end by mid-July, whereas elsewhere the species nest in May–July, as far as is known.

Nectar-feeding Hawaiian honeycreepers
The other subfamily of Hawaiian honeycreepers (Drepanidinae) consists of the mamo, the iiwi, and their allies, seven species in all; of these, only four

▼ *At least eight species of Hawaiian honeycreepers are now extinct, but a few, like this apapane, remain common. Like others in the subfamily Drepanidinae, the sexes are almost identical in appearance.*

P. La Tourrette/Vireo

P. La Tourrette/Vireo

are extant and even they persist in but small numbers or restricted distributions. The members of this group eat nectar, giving rise to the vernacular name "Hawaiian honeycreeper". However, as with "finch", the term "honeycreeper" is meant to convey a lifestyle and associated feeding habits rather than a connection to other birds called honeycreepers, which have evolved independently as nectar-feeders (for example, the *Cyanerpes* honeycreepers of tropical America, in the family Emberizidae). The nectar-feeding Hawaiian honeycreepers nest well above ground, and eat a variety of foods. Several species specialize on flowers and have tongues that are brush-like at their tips, an adaptation for soaking up nectar.

ANTHONY H. BLEDSOE

WINGS WITH TEN PRIMARIES
The three families described below—Estrildidae, Ploceidae, and Passeridae—are small finches of the Old World. Typically they have a cone-shaped bill,

short wings with 10 primaries, and rear their brood in covered nests. They display a remarkable variety of group living, mating systems, and parental care.

ESTRILDID FINCHES
The estrildid finches (family Estrildidae) include five subfamilies: waxbills Estrildinae, grassfinches Poephilinae, mannikins Lonchurinae, and parrot-finches Erythrurinae. They are small birds 9 to 15 centimeters (3½ to 6 inches) long, which feed on small grass seeds, mainly in tropical areas. Most live in Africa, where they are common in grassy areas; several are familiar around villages; and a few feed and nest in forests and swamps. Their cheerful air, bobbing courtship displays, and diet of seeds make them attractive as cage-birds, so the behavior of many species is better known from watching them in captivity than in the field. Males and females are similar in size, and in some species they have similar plumage. The nests are thatched with grass, and often lined with feathers.

Waxbill pairs stay together in a close social

▲ Unusually among birds, many of the Hawaiian honeycreepers, like this iiwi, have kept the original names bestowed upon them centuries ago by the native Hawaiians. Some of these birds played a prominent role in Hawaiian cultural and ceremonial life.

▲ *Three colorful grassfinches and weavers: the Gouldian finch (left) is confined to tropical northern Australia, where its population is steadily declining; the melba finch (center) and the red bishop (right) inhabit African thornveldt and reedbeds respectively.*

bond, perch together in physical contact, preen each other, and rear their young together. One species is the small, red-billed firefinch *Lagonosticta senegala,* which builds its nest of grass and the feathers of domestic or guinea fowl, often in the thatched roofs of African villages, where they are called "animated plums". Adults usually live less than a year and maintain their populations by repeated breeding, rearing several broods in a season. They breed when the rains produce a fresh crop of grass seeds; the adults eat seeds after they fall to the ground, sometimes raiding the raised cones of harvester-termite holes for drying seeds.

The waxbills' nestlings have colorful markings inside the mouth. In the red-billed firefinch the juvenile mouth marking is bright orange-yellow with black spots on the palate and two shades of blue in the light-reflecting tubercles at the corners of the mouth. The bright gape guides the feeding behavior of the parent to the begging young. Other kinds of firefinches—indeed, most other estrildids—each have their own mouth pattern. The patterns may be important to the parents in recognizing their young. The mouth colors of the young disappear when the birds are grown.

Waxbills of the genus *Estrilda* take seeds from the standing stems of grasses, and they nest early in the rains. The melba finch *Pytilia melba* uses its sharp bill to dig into a cover of "runways" on the ground and tree trunks in the dry season so that it can feed on the termites below. One species of

seed-eater, *Pyrenestes ostrinus,* has thick- and thin-billed forms which live together and interbreed in swampy areas of Central Africa. They feed on the seeds of sedges and take the same diet when food is abundant, but in seasons of scarcity the thick-billed form specializes on hard-seeded sedges while the thin-billed form takes other foods.

Mannikins have thicker bills and different courtship displays. The white-rumped munia *Lonchura striata* feeds in irrigated rice fields, eating the rice and the protein-rich filamentous green algae that grows there. Some populations in Malaysia breed twice in a year, each time corresponding with a crop of rice and a bloom of algae, so their breeding seasons are determined by manmade cycles of rice cultivation. Society finches are a domesticated strain of the munia developed centuries ago in China; they are used as foster parents for other estrildid finches and whydahs in captivity, incubating the eggs and rearing the young by regurgitation—although, as the young beg with their heads twisted upside down, society finches are limited as foster parents to birds with this feeding behavior.

In the grassfinch subfamily the best-known species is the zebra finch *Taeniopygia guttata,* which is widespread in dry regions of Australia. It is highly nomadic and breeds in loose colonies after rain. Parrot-finches, 12 species in the genus *Erythrura,* live on tropical islands of the western Pacific, and two species have reached Australia.

They too have bright and unusual colors inside the mouths of the young.

WEAVERS, OR PLOCEID FINCHES

The Ploceidae is a largely African family of 126 species, with a few of those species in Asia and islands of the Indian Ocean. The family is subdivided into buffalo-weavers, sparrow-weavers, true weavers which include bishops and widow finches, and whydahs.

Buffalo-weavers

Buffalo-weavers (subfamily Bubalornithinae) differ from other ploceid finches in that the outer (tenth) primary feather of each wing is large. They are noisy birds which live in flocks and colonies in African savannas (and villages) year-round. One species builds communal nests of thorns and sticks, and the other builds nests of thatched grass. The male black buffalo-weaver *Bubalornis albirostris* is unique among songbirds in having a phallus and in growing a protuberance on the bill in the breeding season.

Sparrow-weavers

The African sparrow-weavers and social weavers (subfamily Plocepasserinae) insert the ends of grass or twigs into a mass of thatch and are "thatchers" rather than "weavers". Several are cooperative breeders. Cooperative colonies of white-browed sparrow-weavers *Plocepasser mahali* last over many generations. The young often remain with their parents and aid them in breeding, or they may move from their natal colony and make a place for themselves in another colony by becoming helpers of the breeding pairs. Sociable weavers *Philetarius socius* build a huge structure of sticks (weighing up to 1000 kilograms, or almost 1 ton) with a common roof and separate nest holes below, each occupied by a pair of breeding birds; the compound nest protects the birds from the extremes of temperature in the Kalahari desert as well as from predators.

Weavers, bishops, and widow finches

Members of the Ploceinae subfamily vary in their behavior. Some are forest-dwelling birds, monogamous, and eat insects. However, most of the 100 species live in grasslands or marshes, are polygynous (one male mating with more than one female), and eat grass seeds, though they may feed insects to their young. Colonies of hundreds of weaver nests are conspicuous in the African landscape. A male builds several nests, attracting several females. He first builds a swing to perch upon, then extends the swing into a ring, and adds a covered basket, keeping the ring as the entrance. He pushes and pulls the nest material into loops and knots around other blades of grass—a process similar to basket-weaving. Some species build a ball-like nest with an entrance hole in one side.

Others finish the nest with a long entrance tube of woven grass. Young males build practice nests which usually fall apart. In the village weaver *Ploceus cucullatus,* males raid each other's nests for green grass. Females are attracted to the male as he hangs upside-down while clinging to the nest, waving his wings in flashes of yellow and black; the female inspects the nest, and if it is to her liking she lays and rears a brood by herself. Females prefer to nest with males in larger colonies. Several weavers nest in mixed-species colonies, where they benefit as in single-species colonies, by seeing other birds returning from new sources of food and by group defence against predators.

The red-billed quelea *Quelea quelea* is a locust-like scourge of cultivated grains. They breed in colonies, a single acacia bush bearing hundreds or thousands of nests, in thousands of bushes. The time when conditions are right for breeding, with seeds in the soft, milky stage suitable for feeding to their young, is short. They breed rapidly: the male builds the nest in two or three days, the female lays three eggs and incubates 10 days, and the young fledge by 10 days and are quickly independent. Flocks follow the seasonal rains over Africa. A bird may breed repeatedly, first where rains fall early, then again hundreds of kilometers away where rains fall a few weeks later.

Among bishops and widow-birds, sexual dimorphism in size and plumage is extreme; males are larger and brightly colored, whereas females and non-breeding males are streaky brown and blend with their grassy habitat. Bishops are named for the hood of red or yellow in the males' breeding

▲ *The male baya weaver builds a nest then coaxes a female to accept it with vigorous wing-flapping displays. As soon as one accepts, moves in and begins a brood, the male proceeds to build another nest close by, repeating the process until he has three or four mates and families.*

▼ *The intensely gregarious zebra finch inhabits desert and grassland across Australia.*

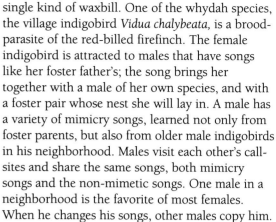

Peter Johnson/NHPA

single kind of waxbill. One of the whydah species, the village indigobird *Vidua chalybeata,* is a brood-parasite of the red-billed firefinch. The female indigobird is attracted to males that have songs like her foster father's; the song brings her together with a male of her own species, and with a foster pair whose nest she will lay in. A male has a variety of mimicry songs, learned not only from foster parents, but also from older male indigobirds in his neighborhood. Males visit each other's call-sites and share the same songs, both mimicry songs and the non-mimetic songs. One male in a neighborhood is the favorite of most females. When he changes his songs, other males copy him.

The young whydahs have the same juvenile mouth markings and begging calls as the nestlings of their foster species, so nestling village indigobirds show the mouth colors of the nestling red-billed firefinches, and the indigobird *V. raricola* has a pink palate and red and light blue tubercles like those of the black-bellied firefinch *Lagonosticta rara.* Juvenile plumage in some whydahs resembles the foster young, especially the unmarked gray of the paradise whydah *Vidua paradisaea* like the melba finch, and the golden plumage of the straw-tailed whydah *V. fischeri* like the purple grenadier *Granatina ianthinogaster.* The African cuckoo-weaver *Anomalospiza imberbis* is also a brood-parasite; it lays its eggs in the nests of grass-warblers, which rear the alien young at the expense of their own brood.

SPARROWS

Old World sparrows (family Passeridae) build a covered nest of thatched grass with a side entrance. They differ from other seed-eating birds by having a vestigial dorsal outer primary feather and an extra bone in the tongue. Several of the 33 species have lived in close association with humans for centuries. One is a highly successful immigrant in areas far from its original range: the house sparrow *Passer domesticus* probably spread from the Middle East along with the movement of agriculture into Europe about 7,000 years ago, then as Europeans colonized other parts of the world in the nineteenth century it went with them. In North America, distinct populations of house sparrows evolved within less than a century and now differ from the founding forms in size and color—those in the northern Great Plains are large, and those in dry areas of the southwest are pale. Different *Passer* species often take over the nests of other birds, like pirates, driving away the nest-builders. The chestnut sparrow *P. eminibey* some-times is a pirate; other times it builds its own nest.

Two other groups of sparrows are rock sparrows (genus *Petronia*) which nest in holes, and snow finches (genus *Montifringilla*) which feed on insects and seeds blown onto high snowfields and nest in rocks above the treeline in Europe and Asia.

ROBERT B. PAYNE

▲ *A long-tailed whydah. These birds use their long tails to attract females in impressive display flights over their grassland territories—the longer the tail, the more successful the display.*

plumage. Some long-tailed species are called "widows" (after the train of tail, as in a widow's veil of mourning), and a few are whydahs (after a Portuguese word with the same meaning). In the long-tailed widow *Euplectes progne,* the male displays with a slow, flapping flight over his grassland territory, tail dangling behind and looking good to females—males in Kenya whose tails were experimentally lengthened by adding parts from other males were able to attract more females to nest. Although it looks like a canary or a weaver, its closest relatives are the *Vidua* finches.

Whydahs, or viduine finches

Whydahs lack a social family life. These brood-parasites (20 species, all in Africa) lay their eggs in the nest of a foster species. The foster parents incubate the eggs and rear the young whydahs along with their own brood. The foster group are waxbills, to which the whydahs are related on the basis of feather arrangement and biochemical genetics, and most whydahs lay in the nests of a

◄ *Native to Southeast Asia, the Java sparrow is widely known as a popular cagebird.*

STARLINGS & THEIR ALLIES

KEY FACTS

ORDER PASSERIFORMES
SUBORDER OSCINES
FAMILY STURNIDAE
• 27 genera • 113 species
FAMILY ORIOLIDAE
• 2 genera • 25 species
FAMILY DICRURIDAE
• 2 genera • 20 species

SIZE

Starlings (Sturnidae)
Total length: 18–45 cm (7–17¾ in)
Weight: 30–105 g (1–3¾ oz)

Orioles (Oriolidae)
Total length: 18–31 cm (7–12¼ in)
Weight: 45–120 g (1½–4¼ oz)

Drongos (Dicruridae)
Total length: 18–38 cm (7–15 in)
Weight: 45–90 g (1½–3 oz)

CONSERVATION WATCH
!!! The critically endangered
species are: Pohnpei mountain
starling *Aplonis pelzeni*; Bali
starling *Leucopsar rothschildi*;
Isabela oriole *Oriolus isabellae*;
Grand Comoro drongo *Dicrurus
fuscipennis*; Mayotte drongo
Dicrurus waldenii.
!! The white-eyed starling *Aplonis
brunneicapilla* is listed as
endangered.
! There are 6 species listed as
vulnerable.

Starlings and mynahs (family Sturnidae), orioles and figbirds (Oriolidae), and drongos (Dicruridae) are Old World birds, with strongest representation in hotter climates. A number of species have glossy black plumage, but many are boldy patterned and brilliantly colored. Most are arboreal, but there are some terrestrial species.

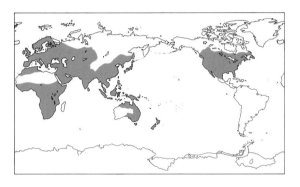

STARLINGS
Almost throughout the English-speaking world the common starling *Sturnus vulgaris* is among the commonest of garden birds. It is a stumpy, rather short-tailed bird with a confident strut and alert, pugnacious manner; its plumage is black, glossed with green and purple, and spangled (in the non-breeding season) with pale brown. Its song, an extraordinary jumble of squeaks, rattles, whistles, and other apparently random sounds, is uttered freely from exposed perches on telephone wires or television aerials. It nests in an untidy jumble of grass and litter stuffed into any available cavity in a tree or a building, and it gathers in large noisy flocks to roost at night.

This bird is in fact a fairly typical member of the Sturnidae, a family of about 113 species found throughout the Old World. The common starling originated in Europe and owes its present more extensive distribution to deliberate introductions by humans, especially during the late nineteenth century when birds were released in North America, South Africa, Australia, New Zealand, and elsewhere, and subsequently prospered.

Some species of starlings are arboreal, fruit-eating birds of jungle and rainforest, but most are insect-eaters inhabiting open country, spending much of their time on the ground. Most species are gregarious, many roost in flocks, and some nest in dense colonies.

There are a number of highly specialized forms with very restricted distributions, but there are also several larger, more significant groups. The genus *Aplonis* comprises about 20 arboreal fruit-eating species found throughout Polynesia and the New Guinea region. About 35 African species, mainly in the genera *Onychognathus, Lamprotornis,* and *Spreo,* include some strikingly beautiful birds with glossy, iridescent plumage, often featuring patches of vivid violet, green, orange, and blue. India is the home of the mynahs (genus *Acridotheres*). Mynahs are mainly dull brown, with patches of naked yellow skin on the head and bold white flashes in the wing. One species, the Indian mynah *Acridotheres tristis,* has been introduced to southern Africa, Australia, Polynesia, and elsewhere; it is a vigorous and aggressive species, often vying with the common starling for domination of the urban environment.

The family also includes the two tickbirds or oxpeckers (genus *Buphagus*), which spend much of their time riding around on buffaloes, rhinos, and other large mammals of the African plains. Researchers are still undecided as to whether the two species of sugarbird (genus *Promerops*) of South Africa are starlings or honeyeaters.

ORIOLES
The typical calls of most orioles have a distinctive timbre—a rich, liquid, bubbling quality that is difficult to describe but easily recognized—and these calls are a characteristic sound of forest and woodland almost throughout the Old World, from Europe to Japan and Australia to southern Africa. The birds themselves are often difficult to observe because they spend most of their lives in dense foliage high in the treetops.

The family has about 25 species in two genera: *Oriolus* (orioles) and *Sphecotheres* (figbirds). Traditionally it has been associated with the starlings and the drongos, but DNA studies suggest a link with the cuckooshrikes (family Campephagidae). Most orioles have a sturdy build with long pointed wings; the bill is long, slender, and slightly de-curved; the eyes are usually red. The males of many species are black and bright yellow, but some are dull green, and streaked below. Females of most species are also dull green and streaked. Most orioles live alone or in pairs. Though they call persistently, they are unobtrusive and deliberate in their general behavior. A few species migrate but most are sedentary. Orioles eat insects and fruit.

One species in this family differs markedly from other orioles in several respects: the figbird

223

◄▲ *The greater raquet-tailed drongo (far left) inhabits India and Southeast Asia; the spatulate, twisted tips to the tail feathers cause a distinctive humming noise in flight. The superb glossy starling (above) is a familiar sight to safari tourists in East Africa, commonly visiting campsites and picnic grounds. The figbird (near left) is common and widespread in Indonesia, New Guinea, and northern and eastern Australia.*

Sphecotheres viridis is gregarious, noisy, active, and quarrelsome. This bird inhabits Indonesia, New Guinea, and northern and eastern Australia, and it is often common in suburban parks and gardens. Its diet consists almost entirely of fruit, especially figs, and a flock can almost strip a tree.

Birds called orioles also occur in North and South America; these species are strikingly similar to Old World orioles in several respects—they are strongly arboreal, have loud ringing songs, build hammock-shaped nests, and the males are boldly patterned in black, yellow, or orange—but they are not related.

DRONGOS

Drongos are bold, aggressive, medium-sized songbirds with short legs, long forked tails, stout bills, an alert upright stance, and flamboyant behavior. The sexes are similar, and the plumage of almost all species is entirely glossy black. Many species have crests, and in several the tail feathers are enormously elongated or otherwise modified

into an ornamental structure. Typical drongo calls include grating metallic notes, modulated whistles, and harsh scolding notes; few have complex or attractive songs. Drongos occur almost throughout the Old World tropics, from Africa to Australia, but the family is best represented in southern Asia. They inhabit wooded country of all kinds, from dense jungle to savanna, but a typical environment is a forest clearing or regrowth at the edge of rainforest. Most species feed primarily on large flying insects captured in flight during sorties from a perch in a tall dead tree or a similar vantage point. Nectar is an important food for some species, and most drongos also capture lizards, small birds, and mammals at least occasionally. They sometimes rob other birds of food in flight. Traditionally the Dicruridae contains about 20 species, but recent DNA studies suggest that the family be expanded to include the monarchs (genus *Monarcha*), fantails (genus *Rhipidura*), magpie-larks (genus *Grallina*), and several other Australasian songbirds.

TERENCE LINDSEY

NEW ZEALAND WATTLEBIRDS

KEY FACTS

ORDER PASSERIFORMES
SUBORDER OSCINES
FAMILY CALLAEATIDAE
• 3 genera • 3 species

SMALLEST & LARGEST

Saddleback *Creadion carunculatus*
Total length: 26 cm (10 in)

Kokako *Callaeas cinerea*
Total length: 38 cm (15 in)

CONSERVATION WATCH
!! The kokako *Callaeas cinerea*
is listed as endangered.
■ The saddleback *Philesturnus
carunculatus* is listed as near
threatened.

C onfined to New Zealand, the Callaeatidae show some similarities to Australian currawongs and apostlebirds, but their ancestry remains uncertain. They may be descendants of an early crow-like stock.

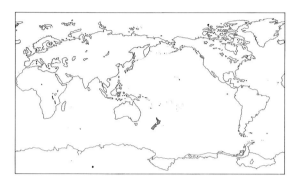

HUIA

The huia *Heteralocha acutirostris* was remarkable for the sexual difference in its bill structure: the female had a very long, slender, and deeply curved bill, while the male's bill was only moderately long and more or less conical. (Among birds, males and females often differ dramatically in plumage and may differ in size, but marked sexual differences in body structure are very rare.) First described in 1835, this bird was known only from dense forests on the North Island of New Zealand, but it has not been reliably reported since 1907 and presumably is extinct.

THE KOKAKO AND THE SADDLEBACK

The huia has two surviving relatives. The kokako *Callaeas cinerea* and the saddleback *Creadion carunculatus,* which together make up the family Callaeatidae, are a very distinct group restricted to New Zealand. They are usually (but tentatively) considered to be most closely related to starlings, bowerbirds, Australian apostlebirds, and mudlarks. All three species are (or were, in the case of the huia) medium-sized songbirds of dense forests, with very reduced powers of flight, strong legs, rounded wings, and conspicuous naked wattles at the gape of the bill. They feed on insects, especially larvae. They are sedentary, long-lived, mate for life, and maintain permanent territories. Even the two surviving members are threatened: they are vulnerable to predation from introduced stoats, weasels, and rats, and to degradation of their environment by the introduced Australian possums.

TERENCE LINDSEY

▲ The huia was remarkable for the difference in bill shape between the sexes (the female's was the longer). Reluctant to fly, it had no defenses against the twin threats of introduced predators and forest destruction, and disappeared by about 1907.

Brian Chudleigh

◄ Formerly widespread on the main islands of New Zealand, the saddleback now survives only on a few small offshore islands.

MAGPIE–LARKS & THEIR ALLIES

ORDER PASSERIFORMES
SUBORDER OSCINES
FAMILY GRALLINIDAE
• 3 genera • 4 species
FAMILY CORCORACIDAE
• 2 genera • 2 species
FAMILY ARTAMIDAE
• 4 genera • 20 species

SMALLEST & LARGEST

Little wood-swallow *Artamus minor*
Total length: 12 cm (4¾ in)
Weight: 15 g (½ oz)

Gray currawong *Strepera versicolor*
Total length: 50 cm (19¾ in)
Weight: 350 g (12½ oz)

CONSERVATION WATCH
■ These species do not appear to be threatened.

A feature of the Australian landscape is the presence of black, or black and white, sociable birds calling to each other. In city parks they could be currawongs or the local magpies. In farmlands they may be any one of a dozen species, for the families in this chapter are common across the whole of Australia.

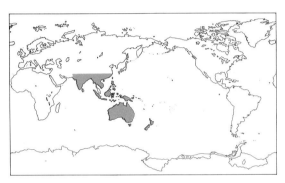

MUD-NEST BUILDERS
The four species grouped under this heading share one characteristic—they all build nests of mud on horizontal branches well off the ground—but they

▼ *Strongly gregarious, the masked wood-swallow is a widespread nomad of the arid interior of Australia.*

R. Drummond

are not as closely related as once thought. Traditionally, the magpie-lark and torrent-lark were put in the family Grallinidae, and the white-winged chough and the apostlebird in the Corcoracidae, but recent DNA studies suggest that the Grallinidae may be part of the Dicruridae (see page 224).

Magpie-lark and torrent-lark
These are thrush-sized birds with glossy black and white plumage, and clear differences between the male and female. The magpie-lark *Grallina cyanoleuca* is common throughout Australia and southern New Guinea in most habitats except forests. Both parents build the nest, incubate the eggs, and brood and feed the nestlings. The young stay with their parents for several weeks after they fledge but join other young birds in late summer. Pairs tend to stay together throughout the year in the same territory, maintaining contact with a piping duet that has given rise to their other common name of "pee-wee". The torrent-lark *G. bruijni* is less well known, as it inhabits the margins of fast-flowing streams in New Guinea's mountains.

White-winged chough and apostlebird
These two species are confined to the shrublands and woodlands of eastern Australia, where they forage on the ground. The white-winged chough *Corcorax melanorhamphus* rakes the litter and probes the ground with its long curved bill in search of insects and their larvae, while the apostlebird *Struthidea cinerea,* with its broad finch-like bill, is mainly a seed-eater. Both species are very sociable and are usually encountered in groups of five to twelve, consisting of the parent birds and their offspring from the previous two or three breeding seasons. It takes several years to acquire the necessary foraging skills and experience to be able to raise a family, and even then pairs are unlikely to be successful unless they have helpers to share the task of provisioning the young. Helpers also share in nest-building, incubation, brooding, and tending fledglings, so breeding is very much a cooperative affair.

WOOD-SWALLOWS

The ten members of the family Artamidae are all in the same genus, *Artamus*. They have blue-gray bills with a black tip, short legs, and long strong wings shaped so that their flight silhouette appears very similar to that of the common starling. The family is largely Australasian·in distribution and they forage in all types of habitat if there are enough insects to make it worthwhile. Although they are primarily insect-eaters, their brush-tongues enable them to lap nectar and pollen when available. Most of the family remain in the same area year-round, but at least three species are widely nomadic and exploit tropical and temperate environments at different times of the year, in large flocks that sometimes contain more than one species. Several species occur throughout islands to the north of Australia, reaching Fiji to the east, and Southeast Asia and India.

When not trawling for aerial insects these very sociable birds frequently perch side by side on conspicuous branches (or telegraph poles) and preen each other; at night they tend to roost together in a swarm clustered on the trunk of a tree or in a hollow. Wood-swallow nests are usually frail structures of fine twigs, and birdwatchers can often see the eggs through the bottom of the nest. In arid regions, nests may be built within six days of rain falling and eggs laid within 12 days, which is considerably shorter than the normal nest building period. Both parents build the nest, incubate the eggs, and feed the young, sometimes assisted by an extra helper.

AUSTRALIAN MAGPIES, OR BELL-MAGPIES

Previously placed in Cracticidae, the 10 species in three genera are now sometimes included with the wood-swallows in the Artamidae. Despite sharing a common name, they are not related to magpies of the family Corvidae (see page 232). They all tend to be black, white, and gray, stockily built, with strong bills usually colored gray with a black tip, and are usually fine singers. The three genera are quite distinct: the *Cracticus* butcher-birds are a little larger than a thrush and are very similar to the shrikes of other continents; the *Gymnorhina* magpies are about twice their size; and the *Strepera* currawongs are the largest, at up to about 50 centimeters (19¾ inches) long.

Butcher-birds and magpies are chiefly found in savanna woodlands and shrublands, whereas the currawongs are mainly birds of the forests. Four species of butcher-birds and a magpie are found in New Guinea but no currawongs. The magpie *G. tibicen* has been introduced to New Zealand, which had no native members of the family. Apart from these examples and the very doubtful inclusion of the Bornean bristlebird *Pityriasis gymnocephala,* of which virtually nothing is known, the family is confined to Australia. Most parts of the country have a couple of species, and they are

among the commonest local birds. Magpies and butcher-birds tend to be resident year-round, often in groups larger than the simple pair because the young stay on long after they become independent. Currawongs tend to wander nomadically through the forests in large flocks after breeding is finished; during the winter they are common sights in many towns in eastern Australia, where their "currawong" call is well known.

All species are basically insect-eaters, but when such food is scarce they will eat seeds and carrion. Butcher-birds generally take their prey in the air or on the ground after flying a swift sortie from a convenient perch. Magpies usually forage as they walk over open ground, probing soft areas of soil and turning over sticks, cow-pats, and other likely hiding places for insects. Currawongs, with the largest, dagger-shaped bill, concentrate on larger prey and have been seen plundering the nests of other birds and even killing small passerines. All members of this family build bulky stick-nests; the female does most of the work and the incubating, but the male and sometimes other group members help to feed the nestlings and fledglings.

IAN ROWLEY

▲ A bold, alert, and aggressive bird, the black currawong is a common panhandler at forest picnic grounds and campsites in the highlands of Tasmania. In winter many form flocks and descend to coastal lowlands. The sexes are similar.

Tom & Pam Gardner

BOWERBIRDS & BIRDS OF PARADISE

ORDER PASSERIFORMES
SUBORDER OSCINES
FAMILY PTILONORHYNCHIDAE
• 8 genera • 20 species
FAMILY PARADISAEIDAE
• 17 genera • 42 species

SMALLEST & LARGEST

BOWERBIRDS (PTILONORHYNCHIDAE)

Golden bowerbird *Prionodura newtoniana*
Total length: 22 cm (8¾ in)
Weight: 70 g (2½ oz)

Great bowerbird *Chlamydera nuchalis*
Total length: 40 cm (15¾ in)
Weight: 230 g (8 oz)

BIRDS OF PARADISE (PARADISAEIDAE)

King bird of paradise *Cincinnurus regius*
Total length: 25 cm (9⅘ in)
Weight: 50 g (1⅘ oz)

Black sicklebill *Epimachus fastuosus*
Total length: 110 cm (43⅓ in)
Weight: 320 g (11⅓ oz)

CONSERVATION WATCH

! There are 8 species listed as vulnerable: Archbold's bowerbird *Archboldia papuensis*; fire-maned bowerbird *Sericulus bakeri*; ribbon-tailed astrapia *Astrapia mayeri*; black sicklebill *Epimachus fastuosus*; Macgregor's bird of paradise *Macgregoria pulchra*; Goldie's bird of paradise *Paradisaea decora*; blue bird of paradise *Paradisaea rudolphi*; Wahnes's parotia *Parotia wahnesi*.

T hese two groups, found only in New Guinea, Australia, and the Moluccas, include some of the most behaviorally complex and bizarrely plumaged birds. Bower-building and decorating by male bowerbirds, and the fantastically ornate plumages and associated displays of male birds of paradise, have long attracted the attention of ornithologists all over the world. The two groups were once considered closely related, forming one family, but recent genetic studies have shown that they are quite separate.

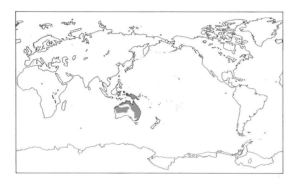

BOWERBIRDS
Bowerbirds (family Ptilonorhynchidae) are closely related to Australian lyrebirds, scrub-birds, and, surprisingly, treecreepers. They are stout, heavy-billed, strong-footed birds. Nine species are peculiar

to New Guinea, where the group presumably originated, eight are found only in Australia, and two occur in both areas. The three species that are called catbirds, because of their cat-like call, are monogamous, the sexes forming pairs to defend a territory and care for offspring. All 17 other species of bowerbirds are polygynous; the females build nests, incubate eggs, and raise young alone, and are appropriately drably colored in browns, grays, or greens. Both sexes in most of the polygynous species mimic other bird calls, especially those of predatory birds, and other sounds.

Bowerbirds live in rainforest or other wet forests, except the five avenue-bower-building *Chlamydera* species, which inhabit dry woodlands, grasslands, and, in three cases, the arid Australian interior.

Bowerbirds eat mostly fruit, but nestlings are

▶ An adult male satin bowerbird rearranges his treasure, displaying his species' strong bias toward blue. As in other bowerbirds, the display area is decorated with a variety of small objects, often including plastic drinking straws and similar picnic litter.

Frithfoto

fed differing proportions of animals, usually insects; catbird nestlings are often fed other birds' nestlings. Some species also eat flowers, flower and leaf buds, fresh leaves, and succulent vine stems.

Nests are bulky cups of leaves, fern fronds, and vine tendrils on an untidy stick foundation built in a shrub, tree fork, or vine tangle, except for the golden bowerbird *Prionodura newtoniana,* which builds its nest in a tree crevice. One to three eggs are laid; incubation lasts from 19 to 27 days, and chicks fledge 18 to 32 days after hatching.

Recent genetic studies support inclusion of the enigmatic and probably extinct pio pio *Turnagra capensis* of New Zealand with the bowerbirds.

BIRDS OF PARADISE

Generally stout- or long-billed and strong-footed, birds of paradise are crow-like in appearance and size; genetically the family Paradisaeidae is closest to crows and their allies. The group is divided into the subfamilies Paradisaeinae, with 38 typical birds of paradise, and Cnemophilinae, with three little-known species—Loria's *Cnemophilus loriae,* yellow-breasted *Loboparadisea sericea,* and crested *Cnemophilus macgregorii* birds of paradise. They live in rainforest, moss forest, or swamp forest and may visit nearby gardens. Fruit dominates their diet.

Most of them live in mountainous New Guinea and immediately adjacent islands. The exceptions are the paradise crow *Lycocorax pyrrhopterus* and the standardwing bird of paradise *Semioptera wallacii* on the Moluccan islands, and the paradise riflebird *Ptiloris paradiseus* and Victoria's riflebird

P. victoriae in eastern Australia. The magnificent riflebird *P. magnificus* and trumpet manucode *Manucodia keraudrenii* extend their New Guinea ranges to include the rainforests of Cape York Peninsula, in northeastern Australia.

Of the typical birds of paradise in which the sexes are almost identical (monomorphic), several species have been found to be monogamous—the male and female sharing nesting duties. As a result the monomorphic paradise crow and two paradigallas have also been considered monogamous. However, I recently studied a nesting short-tailed paradigalla *Paradigalla brevicauda* and observed only a single parent, presumably the female. Of the remaining 30 typical birds of paradise which are sexually dimorphic (sexes unalike), almost half are known to be polygynous—one male fertilizing several females, and the females raising their young alone. The remainder are assumed to be likewise.

Nests are bulky cups of leaves, ferns, orchid stems, and vine tendrils placed in a tree fork. Few include sticks, except that of the crested bird of paradise which is a domed mossy nest atop a few sticks. This species was considered to be a bower-bird, and to be monogamous, but I recently studied it nesting and confirmed that it is a bird of paradise, with a lone (presumably female) nesting parent. The king bird of paradise *Cincinnurus regius* nests in a tree hollow. Birds of paradise lay one to three eggs, beautifully marked and colored, and these take between 16 and 23 days to hatch. Nestlings leave the nest when they are 14 to 30 days old.

CLIFFORD B. FRITH

▲ *The emperor bird of paradise is restricted to the mountains of the Huon Peninsula, Papua New Guinea.*

▼ *The regent bowerbird inhabits the forests of central-eastern Australia.*

Tom & Pam Gardner

► A male Victoria's riflebird at the climax of his extraordinary display. Preferred display perches are usually broken-off tree stubs several meters from the ground in dense rainforest.

D. Parer & E. Parer-Cook/Auscape International

▲ Most members of the genus Paradisaea, like this raggiana bird of paradise, congregate to display in leks: a tree full of these birds all in simultaneous display, quivering their long lacy flank plumes and calling hysterically, is one of the most spectacular sights among birds.

Hans & Judy Beste/Auscape International

SPECTACULAR MATING RITUALS

Birds of paradise are well known for the males' ornate plumage and fantastic courtship displays. Males of the various polygynous species court females in differing ways, but the best known are species in which males congregate to display in groups, known as leks. These include all but one of the plumed or "true" birds of paradise of the genus *Paradisaea*. The males call from and hop about lek perches and spread raised plumes while posturing and performing dances in unison. Studies of raggiana bird of paradise *P. raggiana* leks revealed that most visiting females mate with the same individual male, presumably the dominant and most fit on the lek. In this way, females optimize the quality of male genes for their offspring.

True lekking behavior is known also in the standardwing bird of paradise from the Moluccas, and in Stephanie's astrapia *Astrapia stephaniae* of New Guinea.

Males of some other conspicuously sexually dimorphic (sexes unalike) species do not congregate, but display alone; for example, in the superb bird of paradise *Lophorina superba* and the riflebirds, a solitary male displays on a tree stump, fallen tree trunk, or perch. Male parotias (genus *Parotia*) perform elaborate dances on cleared ground courts and may form dispersed leks.

In the polygynous bowerbirds, the male constructs and decorates a bower of grasses, ferns, orchids, or sticks, and there he calls and displays to attract and impress females. In some adult male bowerbirds, plumage is spectacularly colorful, but in others it is uniformly dull like that of the female; males with colorful plumage build simple bowers, whereas unadorned males build complex ones. This clearly indicates that bowers represent a transfer of sexual attraction from the male's appearance to a structure and its decorations. Males go so far as to steal decorations from one another's bowers. Studies of the satin bowerbird *Ptilonorhynchus violaceus* reveal that females choose to mate with males at larger, better-decorated bowers.

There are four bower types: court, mat, maypole, and avenue. Only the tooth-billed bowerbird *Scenopoeetes dentirostris* makes a "court", by clearing a patch of forest floor and decorating it with upturned fresh leaves. The only mat bower is that of the rare Archbold's bowerbird *Archboldia papuensis,* a large black bird of highland New Guinea, which accumulates a mat of fern fronds on the forest floor and decorates it with snail shells, beetle wing cases, fungus, charcoal, and other items, and drapes orchid stems on the perches above. Maypole bowers—stick structures built around one or several sapling stems—are constructed by the four New Guinea gardener bowerbirds (genus *Amblyornis)* and the golden bowerbird *Prionodura newtoniana* of northeast Australian rainforests. The remaining eight species make avenue bowers—vertical stick-walls standing in the ground—in simple or complex forms. Satin bowerbird bowers have been known to exist at one site for 50 years.

Jean-Paul Ferrero/Auscape International

▲ Smallest and most vivid of its family, the king bird of paradise is common, noisy, but difficult to observe in the dense thickets it prefers. Though each male displays independently there are usually other males not too far away.

Jim Frazier/Mantis Wildlife Films

◄ A male great bowerbird coaxes a female to his display area. This particular bower is unusual in being roofed; most are left open at the top.

CROWS & JAYS

KEY FACTS

ORDER PASSERIFORMES
SUBORDER OSCINES
FAMILY CORVIDAE
• *c.* 25 genera • *c.* 117 species

SMALLEST & LARGEST

Hume's ground jay *Pseudopodoces humilis*
Total length: 20 cm (8 in)

Raven *Corvus corax*
Total length: 66 cm (26 in)

CONSERVATION WATCH

!!! The Hawaiian crow *Corvus hawaiiensis* and Mariana crow *Corvus kubaryi* are critically endangered.
!! The white-throated jay *Cyanolyca mirabilis* and dwarf jay *Cyanolyca nana* are listed as endangered.
! There are 9 vulnerable species, including: Flores crow *Corvus florensis;* hooded treepie *Crypsirina cucullata;* Sichuan jay *Perisoreus internigrans;* Ethiopian bush-crow *Zavattariornis stresemanni.*

▼ *An Australian raven utters its harsh unmusical call.*

The crow family Corvidae is a very successful branch of passerines, which probably originated in Australia when it was isolated from the Asian landmass. Only when the continents drifted closer together, about 20 to 30 million years ago, could landbirds cross the water and an exchange of species take place. Once the original corvids reached Asia, extensive evolution seems to have occurred, resulting in the many different forms that have spread to all parts of the world. Approximately 117 species may be broadly grouped into crows (including jackdaws, rooks, and ravens), choughs, nutcrackers, typical magpies, blue-green magpies, treepies, typical jays, American jays, gray jays, and ground jays.

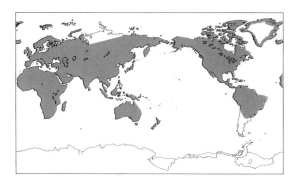

LONG-LEGGED FORAGERS

Most corvids are medium to large in size, have nostrils covered with bristles, and relatively long legs with characteristic scaling. Their color varies from the somber black of the raven through to the brilliant reds and greens of the Asian magpies. Both parents build the nest and feed the young, but only the female incubates the eggs and broods the nestlings; during that time she is fed by the male. Many corvids hide surplus food in caches which they relocate and use long afterwards.

Members of the family are mainly found in forests, open woodland, scrubland, and plains. Their long legs enable them to walk and hop over the ground quickly, foraging as they go, probing with their strong bills to search for insects, tear meat from carcasses, harvest berries, or pick up fallen seeds. Many of the smaller species forage extensively in the forest canopy and some specialize in harvesting particular foods such as nuts and pine seeds, which they may cache. Many corvids do not breed until they are at least two years old, and in the meantime the immature birds form large flocks that feed, fly, and roost together, exploiting the food available over a wide area, sometimes becoming pests. In many species, a breeding pair tends to be resident in one area or territory with the same partner for as long as they both survive, and they defend that area against intrusion by others of the same species. Other species, such as the rook *Corvus frugilegus,* jackdaw *C. monedula,* and pinyon jay *Gymnorhinus cyanocephala,* prefer to nest colonially. A few species, such as the Florida scrub jay *Aphelocoma coerulescens* and the Mexican jay *A. ultramarina,* have adopted a cooperative way of life and live in social groups consisting of the progeny from previous years, which stay in the family long after they have reached sexual maturity, helping the breeders to raise the current crop of young.

CORVID GROUPS

Crows, jackdaws, rooks, and ravens (43 species in the genus *Corvus*) vary in size from 33 to 66 centimeters (13 to 26 inches) and are basically black, although a few species show areas of white, gray, or brown. This genus has representatives in every continent except South America, with several living in close proximity to humans; for example, I have seen the hooded crow *C. corone cornix* fighting its reflection in the gilded domes of the Kremlin and the house crow *C. splendens* peering into bicycle baskets in Mombasa.

C.A. Henley

The two black choughs (genus *Pyrrhocorax*) closely resemble the true crows but have finer bills, one scarlet and one yellow. Both live in rocky environments in Eurasia, either on coastal cliffs or in mountains where they live above the treeline for most of the year, eating mainly insects and berries.

In contrast the two nutcrackers (genus *Nucifraga*) are usually permanent residents of conifer-clad mountains, one in America and the other in Europe and Asia; they are remarkable for the extent of their dependence on cached stores of pine seeds to last them through hard winters. The distinct, but very similar gray jays (genus *Perisoreus*) have much the same distribution to that of the nutcrackers; they too are permanent residents of coniferous forests, one species in North America and the other in Eurasia and Asia.

The true magpie *Pica pica* is common throughout most of North America, Europe, and Asia; it is easily identified by its black and white plumage, long tail, and large family groups. Magpies like thick shrubby cover for their roofed nests, but they forage for insects and seeds in the open, especially in agricultural lands and, more recently, suburbia. They occasionally rob other birds' nests of eggs and young. In California a second species, *P. nuttalli,* is clearly recognized by its yellow bill. The azure-winged magpie *Cyanopica cyana* of Spain and eastern Asia is a near relative with a similar lifestyle.

The blue, green, and Whitehead's magpies (genus *Cissa*) are about the same size as true magpies but have even longer tails. The blue magpies are birds of thickets and woodlands occurring throughout the Indian subcontinent, and from Southeast Asia to Taiwan and to the larger Indonesian islands of Java and Borneo. In this region the green and chestnut-red species of this genus tend to live in jungle and forest.

The 10 treepies (genera *Crypsirina* and *Dendrocitta)* share much the same distribution as the *Cissa* magpies but are more plainly colored in shades of gray, black, and white; they have black feet, strongly curved black bills, and long tails. They spend most of their time in trees, often in mixed-species flocks of foraging birds, especially drongos.

True jays (three species in the genus *Garrulus*) have unusual blue and black barred feathers, and are confined to Europe and Asia, where they eat a wide range of foods. They are especially fond of acorns, which they may cache. The blue or American jays are very varied and many have strikingly beautiful plumage. Although the 30 or so species are placed in 7 different genera, they are thought to have originated from the same stock and occur throughout the Americas from Canada to Argentina, in scrub, woodland, and forest. Most feed mainly on insects and fruits, but some have become extremely specialized. For example, wherever the piñon pine grows in western North America it provides the pinyon jay with a staple

Wilhelm Möller/Ardea Photographics

▲ Both sexes of the Eurasian jay cooperate in raising their brood of young. Normally noisy and demonstrative birds, jays are extremely quiet and inconspicuous when nesting.

John Cancalosi/Auscape International

◀ Noisy, lively, and inquisitive, the green jay often visits farmyards, ranches, and suburban gardens of the more arid parts of Central America. It favors bushy thickets and streamside vegetation.

diet of seeds; the birds cache the seeds to provide a reserve for the breeding season.

Ground jays inhabit the high deserts of Central Asia. Four species (genus *Podoces*) are much the same size as the other jays, but the fifth, Hume's ground jay *Pseudopodoces humilis,* is sparrow-sized.

Finally, three birds are currently included in the Corvidae for want of a better place to put them. These are the Ethiopian bush-crow *Zavattariornis stresemanni;* the piapiac *Ptilostomus afer* of West and Central Africa, and the crested jay *Platylophus galericulatus* of Malaysia and Indonesia.

IAN ROWLEY

FURTHER READING

- **INTRODUCING BIRDS (P. 14)**

Gill, F.B., 1995. *Ornithology,* 2nd edn. W.H. Freeman, New York.

Farner, D.S., J.R. King & K.C. Parkes (eds), 1971–1993. *Avian Biology, vols I–IX.* Academic Press, New York.

Pettingill, O.S., Jr, 1985. *Ornithology in Laboratory and Field.* Academic Press, New York.

Proctor, N.S., & P.J. Lynch, 1993. *Manual of Ornithology: Avian Structure and Function.* Yale University Press, New Haven.

Welty, J.C., & L.F. Baptista, 1990. *The Life of Birds.* Saunders College Publishers, Fort Worth.

- **CLASSIFYING BIRDS (P. 19)**

Bock, W.J., 1994. 'Nomenclatural history of avian family-group names and their synonyms', *Bull. Amer. Mus. Nat. Hist.,* 222: 1-281; 1995, Erratum, ibid., 222: 1-3.

Mayr, E., & W.J. Bock, 1994. 'Provisional classifications versus standard avian sequences', *Ibis,* 136: 12-18.

Sibley, C.G., & B.L. Monroe, Jr, 1990. *Distribution and taxonomy of Birds of the World.* Yale University Press, New Haven.

- **BIRDS THROUGH THE AGES (P. 23)**

Chatterjee, S., 1997. *The Rise of Birds: 225 Million Years of Bird Evolution.* Johns Hopkins University Press, Baltimore.

Chiappe, L., 1995. 'The first 85 million years of avian evolution', *Nature,* London, 378: 349-355.

Feduccia, A., 1996. *The Origin and Evolution of Birds.* Yale University Press, New Haven.

Martin, L.D., 1983. 'The origin and early radiation of birds', pp. 291-338 in A.H. Brush & G.A. Clark, Jr, (eds). *Perspectives in Ornithology.* Cambridge University Press, Cambridge.

Olson, S.L., 1985. 'The fossil record of birds', pp. 79-238 in D.S. Farner *et al. Avian Biology,* vol. 8. Academic Press, New York.

- **HABITATS & ADAPTATIONS (P. 27)**

Cody, M.L., 1985. *Habitat Selection in Birds.* Academic Press, New York.

Ford, H.A., 1989. *Ecology of Birds: an Australian Perspective.* Surrey Beatty, Chipping Norton, NSW.

Fuller, R.J., 1982. *Bird Habitats in Britain.* Poyser, Calton, UK.

Lack, D., 1971. *Ecological Isolation in Birds.* Blackwell, Oxford.

Perrins, C.M., & T.R. Birkhead, 1983. *Avian Ecology.* Blackie, Glasgow.

Weiner, J., 1994. *The Beak of the Finch: a Story of Evolution in Our Time.* Vintage, London.

- **BIRD BEHAVIOR (P. 32)**

Catchpole, C.K., & P.J.B. Slater, 1995. *Bird Song: Biological Themes and Variations.* Cambridge University Press.

Ficken, M.S., 1977. 'Avian Play', *Auk,* 94: 573-82.

Nice, M.M., 1962. 'Development of behaviour in precocial birds', *Trans. Linn. Soc.,* New York, 8: 1–211

- **ENDANGERED SPECIES (P. 37)**

Collar, N.J., M.J. Crosby & A.J. Stattersfield, 1994. *Birds to Watch 2: The World List of Threatened Birds.* Birdlife Conservation Series, no. 4. 407pp.

Greenway, J.C., Jr, 1967. *Extinct and Vanishing Birds of the World,* 2nd edn. Dover Publications, New York.

- **KINDS OF BIRDS (PP. 44–233)**

del Hoyo, J., A. Elliott & J. Sargatal (eds), 1992-1997. *Handbook of the Birds of the World. Volumes 1-4.* Lynx Edicions/ICBP, Barcelona. Further volumes to be published.

- **RATITES & TINAMOUS (P. 46)**

Blake, E.R., 1977. *Manual of Neotropical Birds, Vol. I.* Chicago University Press, Chicago. pp. 8-80.

Davies, S.J.J.F., 1976. 'The natural history of the Emu in comparison with that of other ratites', *Proc. XIV Int. Orn. Congr.,* pp 109-20.

Sauer, E.G.F., & E.M. Sauer, 1966. 'The behaviour and ecology of the South African Ostrich', *Living Bird,* 5: 45-75.

- **ALBATROSSES & PETRELS (P. 50)**

Alexander, W.B., *Birds of the Ocean,* 2nd edn. Putnam, London.

Harrison, P., 1983. *Seabirds: An Identification Guide.* Houghton Mifflin, Boston.

Warham, J., 1990. *The Petrels: Their Ecology and Breeding Systems.* Academic Press, London.

Warham, J., 1996. *The Behaviour, Population Biology and Physiology of the Petrels.* Academic Press, London.

- **PENGUINS (P. 55)**

Williams, T.D., 1995. *The Penguins.* Oxford University Press, Oxford.

- **DIVERS & GREBES (P. 58)**

Fjeldså, J., 1973. 'Antagonistic and heterosexual behavior of the Horned Grebe *Podiceps auritus*', *Sterna,* 12: 161-217.

Fjeldså, J., 1981. 'Comparative ecology of Peruvian Grebes. A study in mechanisms for evolution of ecological isolation, Vidensk', *Meddr. dansk, naturh. Foren.,* 144: 125-246.

Johnsgard, P.A., 1987. *Diving Birds of North America.* University of Nebraska Press, Lincoln.

- **PELICANS & THEIR ALLIES (P. 61)**

Harrison, P., 1983. *Seabirds: An Identification Guide.* Houghton Mifflin, Boston.

Johnsgard, P.A., 1993. *Cormorants, Darters and Pelicans of the World.* Smithsonian Institution Press, Washington.

Nelson, J.B., 1978. *The Sulidae: Gannets and Boobies.* Oxford University Press, Oxford.

van Tets, G.F., 1965. 'A comparative study of some social communication patterns in the Pelecaniformes', *Orn, Monogr.* no. 2.

- **HERONS & THEIR ALLIES (P. 67)**

Hancock, J.A., & H. Elliot, 1978. *The Herons of the World.* London Editions, London.

Hancock, J.A., J.A. Kushlan & M.P. Kahl, 1992. *Storks, Ibises and Spoonbills of the World.* Academic Press, London.

Kear, J., & N. Duplaix-Hall, 1975. *Flamingos.* T & A.D. Poyser, Berkhamsted.

- **RAPTORS (P.75)**

Brown, L., & D. Amadon, 1989. *Eagles, Hawks and Falcons of the World.* Spring Books, London.

Cade, T., 1982. *The Falcons of the World.* Collins, London.

Mundy, P., *et al.,* 1992. *The Vultures of Africa.* Academic Press, London.

Newton, I., & P. Olsen (eds), 1990. *Birds of Prey.* IPC, Sydney.

Olsen, P., 1995. *Australian Birds of Prey.* Johns Hopkins, Baltimore.

- **WATERFOWL & SCREAMERS (P. 81)**

Johnsgard, P., 1978. *Ducks, Geese and Swans of the World.* University of Nebraska Press, Lincoln, Nebraska.

Madge, S., & H. Burn, 1988. *Wildfowl (An Identification Guide to the Ducks, Geese and Swans of the World).* Christopher Helm, London.

Marchant, S., & P. Higgins (eds), 1990. *Handbook of Australian, New Zealand, and Antarctic Birds. Vol. 1 (Ratites to Ducks).* Oxford Univ. Press, Melbourne.

Todd, F.S., 1996. *Natural History of the Waterfowl.* Ibis Publishing Co., Vista, California.

- **GAMEBIRDS & THE HOATZIN (P. 88)**

Delacour, J., & D. Amadon, 1973. *Curassows and Related Birds.* American Museum of Natural History, New York.

Johnsgard, P.A., 1983. *The Grouse of the World.* Croom Helm, London.

Johnsgard, P.A., 1988. *The Quails, Partridges, and Francolins of the World.* Oxford University Press, London.

Jones, D.N., R.W.R.J. Dekker & C.S. Roselaar, 1995. *The Megapodes: Megapodiidae.* Oxford University Press, London.

- **CRANES & THEIR ALLIES (P. 95)**

Ellis, D.H., G.F. Gee & C.M. Mirande (eds), 1996. *Cranes: Their Biology, Husbandry, and Conservation.* Hancock House, Blaine.

Johnsgard, P.A., 1991. *Bustards, Hemipodes, and Sandgrouse: Birds of Dry Places.* Oxford University Press, London.

Johnsgard, P.A., 1983. *Cranes of the World.* Indiana University Press, Bloomington.

Ripley, S.D., 1977. *Rails of the World: A Monograph of the Family Rallidae.* M.F. Feheley Publishers Limited, Toronto.

- **WADERS & SHOREBIRDS (P. 102)**

Harrison, P., 1983. *Seabirds: An Identification Guide.* Houghton Mifflin, Boston.

Hayman, P., J. Marchant & T. Prater, 1986. *Shorebirds: An Identification Guide.* Houghton Mifflin, Boston.

- **PIGEONS & SANDGROUSE (P. 114)**

Frith, H.J., 1982. *Pigeons and Doves of Australia.* Rigby, Adelaide.

Goodwin, D., 1970. *Pigeons and Doves of the World,* 2nd edn. British Museum (Natural History), London.

Johnsgard, P.A., 1991. *Bustards, Hemipodes, and Sandgrouse: Birds of Dry Places.* Oxford University Press, London.

- **PARROTS (P. 118)**

Beissinger, S.R, & N.F.R. Snyder, 1992. *New World Parrots in Crisis: Solutions from Conservation Biology.* Smithsonian Institution Press, Washington.

Forshaw, J.M., 1989. *Parrots of the World,* 3rd (revised) edn. Lansdowne Editions, Willoughby, NSW.

Juniper, T., & M. Parr, 1997. *Parrots: A Guide to the Parrots of the World.* Pica Press, Mountfield.

Smith, G.A., 1975. 'Systematics of parrots', *Ibis,* 117: 18-68.

- **TURACOS & CUCKOOS (P. 125)**

Baker, E.C.S., 1942. *Cuckoo Problems.* Witherby, London.

Chance, E., 1922. *The Cuckoo's Secret.* Sidgwick & Jackson, London.

Fry, C.H., S. Keith & E.K. Urban, 1988. *The Birds of Africa, vol. 3.* Academic Press, London.

Rowan, M.K., 1983. *The Doves, Parrots, Louries and Cuckoos of Southern Africa.* Croom Helm, Beckenham, Kent.

Wyllie, I. 1981. *The Cuckoo.* Batsford. London.

- **OWLS, FROGMOUTHS & NIGHTJARS (P. 128)**

Burton, J.A., (ed.), 1992. *Owls of the World: Their Evolution, Structure and Biology.* Lowe, London.

Hollands, D., 1995. *Birds of the Night.* Reed, Sydney.

Mikkola, E., 1982. *Owls of Europe.* Poyser, Calton, UK.

- **SWIFTS & HUMMINGBIRDS (P. 134)**

Chantler, P., & G. Driessens, 1995. *Swifts: A Guide to the Swifts and Treeswifts of the World.* Pica Press, Sussex.

Johnsgard, P.A., 1997. *The Hummingbirds of North America,* 2nd edn. Smithsonian Institution, Washington D.C.

Skutch, A., 1973. *Life of the Hummingbird.* Vineyard Books, New York.

- **MOUSEBIRDS & TROGONS (P. 138)**

Cunningham-van Someren, G.R., & C.H. Fry, 1988, pp. 255-63 in C.H. Fry, S. Keith & E.K. Urban (eds). *The Birds of Africa, vol. III.* Academic Press, London.

Decoux, J.P., 1988, pp. 247-54 in C.H. Fry, S. Keith & E.K. Urban (eds). *The Birds of Africa, vol. III.* Academic Press, London.

Skutch, A.F., 1944. 'Life History of the Quetzal', *Condor,* 46: 213-35.

Wheelwright, N.T., 1983. 'Fruits and ecology of Resplendent Quetzals', *Auk,* 100: 286–301.

- **KINGFISHERS & THEIR ALLIES (P. 140)**

Forshaw, J.M., 1983–1994. *Kingfishers and Related Birds: Part I, Alcedinidae; Part II, Todidae to Phoeniculidae; Part III, Bucerotidae.* Lansdowne Editions, Melbourne.

Fry, C.H., K. Fry & A. Harris, 1992. *Kingfishers, Bee-eaters and Rollers.* Helm, London.

Kemp, A.C., 1995. *The Hornbills: Bucerotiformes.* Oxford University Press, Oxford.

- **WOODPECKERS & BARBETS (P. 152)**

Short, L.L., 1982. *Woodpeckers of the World.* Delaware Museum of Natural History, Delaware.

Skutch, A.F., 1971. 'Life history of the Keel-billed Toucan', *Auk,* 88: 381-96.

Winkler, H., D.A. Christie & D. Nurney, 1995. *Woodpeckers: A Guide to the Woodpeckers, Piculets and Wrynecks of the World*. Pica Press, Sussex.

• BROADBILLS & PITTAS (P. 158)

Johnsgard, P.A., 1993. *Cormorants, Darters, and Pelicans of the World*. Smithsonian Institution Press, Washington D.C.

Lambert, F., & M. Woodcock, 1996. *Pittas, Broadbills & Asities*. Pica Press, Sussex.

Olson, S., 1971. 'Taxonomic comments on the Eurylaimidae', *Ibis, 113*: 507-16.

• OVENBIRDS & THEIR ALLIES (P. 162)

Ridgely, R.S., & G. Tudor, 1994. *The Birds of South America. Vol. 2: The Suboscine Passerines*. University of Texas Press, Austin.

Willis, E.O., 1967. 'The behavior of Bicolored Antbirds', *University of California Publ. Zool., 79*: 1-132.

Willis, E.O., 1972. 'The behavior of Spotted Antbirds', *A.O.U. Mongr., 10*: 1-162.

• TYRANT FLYCATCHERS & THEIR ALLIES (P. 166)

Ridgely, R.S., & G. Tudor, 1994. *The Birds of South America. Vol. 2: The Suboscine Passerines*. University of Texas Press, Austin.

Sick, H., 1967. 'Courtship behavior in manakins (Pipridae): a review', *Living Birds, 6*: 5-22.

Snow, D.W., 1982. *The Cotingas*. British Museum (Natural History), London.

Traylor, M.A., Jr, & J.W. Fitzpatrick, 1982. 'A survey of tyrant flycatchers', *Living Bird, 19*: 7-50.

• LYREBIRDS & SCRUB-BIRDS (P. 169)

Robinson, F.N., & H.S. Curtis, 1996. 'The vocal displays of the Lyrebirds: Menuridae', *Emu, 96*: 258-275.

Robinson, F.N., & H.J. Frith, 1981. 'The Superb Lyrebird *Menura novaehollandiae* at Tidbinbilla, ACT', *Emu, 81*: 145-157.

Smith, G.T., 1996. 'Habitat use and management for the Noisy Scrub-bird *Atrichornis clamosus*', *Bird Conservation International, 6*: 33-48.

• LARKS & WAGTAILS (P. 171)

Hall, B.P., & R.E. Moreau, 1970. *An Atlas of Speciation in African Passerine Birds*. British Museum (Natural History), London.

Keith, S., E.K. Urban & C.H. Fry, 1992. *The Birds of Africa. Vol. 4*. Academic Press, London.

Liversidge, R., 1996. 'A new species of pipit in southern Africa', *Bull. Brit. Orn. Cl., 116* : 211-215.

Simms, E., 1992. *British Larks, Pipits and Wagtails*. Harper Collins, London.

• SWALLOWS (P. 174)

Dorst, J., 1962. *The Migrations of Birds*. Heinemann, London.

Turner, A., & C. Rose, 1989. *A Handbook to the Swallows and Martins of the World*. Christopher Helm, London.

Vaurie, C., 1959. *The Birds of the Palearctic Fauna*. Witherby, London.

• CUCKOOSHRIKES (P. 176)

Coates, B.J., 1990. *The Birds of Papua New Guinea. Vol. 2*. Dove Publications, Alderley, Australia.

Collar, N.J., & S.N. Stuart, 1985. *Threatened Birds of Africa and related islands*. ICBP, Cambridge.

Hall, B.P., & R.E. Moreau, 1970. *An Atlas of Speciation in African Passerine Birds*. British Museum (Natural History), London.

• BULBULS & LEAFBIRDS (P. 177)

Ali, S., 1969. *Birds of Kerala*. Oxford University Press, London.

Diamond, A.W., & T.E. Lovejoy (eds), 1985. *Conservation of tropical forest birds: proceedings of a workshop and symposium held at the Eighteenth World Conference of the International Council for Bird Preservation*. ICBP technical publication 4, Cambridge.

Hall, B.P., & R.E. Moreau, 1970. *An Atlas of Speciation in African Passerine Birds*. British Museum (Natural History), London.

• SHRIKES & VANGAS (P. 180)

Dee, T.J., 1986. *The Status and Distribution of the Endemic Birds of Madagascar*. ICBP, Cambridge.

Langrand, O., 1990. *Guide to the Birds of Madagascar*. Yale University Press, New Haven.

Lefranc, N., & T. Worfolk, 1997. *Shrikes: A Guide to the Shrikes of the World*. Pica Press, Mountfield.

Smith, E.F.G., P. Arctander, J. Fjeldså & O.G. Amir, 1991. 'A new species of shrike (Laniidae: Laniarius) from Somalia, verified by DNA sequence data from the only known individual', *Ibis, 133*: 227-235.

• WAXWINGS & THEIR ALLIES (P. 182)

Cramp, S., (ed.), 1988. *Handbook of the Birds of Europe, the Middle East and North Africa. The Birds of the Western Palearctic. Vol. 5. Tyrant Flycatchers to Thrushes*. Oxford University Press, Oxford.

Putnam, L.S., 1949. 'The life history of the Cedar Waxwing', *Wilson Bull., 61*: 141-82.

Skutch, A.F., 1965. 'Life history of the Long-tailed Silky-flycatcher, with notes on related species', *Auk, 82*: 375-426.

• MOCKINGBIRDS & ACCENTORS (P. 183)

Birkhead, M.E., 1981. 'The social behaviour of the Dunnock', *Ibis, 123*: 75-84.

Michener, H., & J.R. Michener, 1935. 'Mockingbirds, their territories and individualities', *Condor, 37*: 97-140.

• DIPPERS & THRUSHES (P. 185)

Bent, A.C., 1964. *Life histories of North American thrushes, kinglets and their allies*. Dover Publications, New York.

Shaw, G., 1978. 'The breeding biology of the Dipper', *Bird Study, 25*: 140-60.

Tyler, S., & S. Ormerod, 1994. *The Dippers*. T. & A.D. Poyser, Calton, UK.

• BABBLERS & WRENS (P. 188)

Blakers, M.D., & P.N. Reilly, 1984. *The Atlas of Australian Birds*. Melbourne University Press, Melbourne. pp. 413-426.

Boles, W.E., 1988. *The Robins and Flycatchers of Australia*. Angus & Robertson, Sydney. pp. 49-75.

Gaston, A.J., 1978. 'Demography of the Jungle Babbler *Turdoides striatus*', *J. Anim. Ecol., 47*: 845-70.

WARBLERS & FLYCATCHERS (P. 192)

Cramp, S., & D.J. Brooks (eds), 1992. *Handbook of the Birds of Europe, the Middle East and North Africa. The Birds of the Western Palearctic. Vol. 6. Warblers*. Oxford University Press, Oxford.

Dunn, J., & K. Garrett, 1997. *A Field Guide to the Warblers of North America*. Houghton Mifflin & Co., Boston.

McGill, A.R., 1970. *Australian Warblers*. The Bird Observers Club, Melbourne. 147 pp.

Parmenter, T., & C. Byers, 1991. *A Guide to the Warblers of the Western Palaearctic*. Bruce Coleman, Uxbridge.

• FAIRY WRENS & THEIR ALLIES (P. 196)

Rowley, I., & E. Russell, 1997. *Fairy-wrens and grasswrens*. Oxford University Press, Oxford. 274 pp.

Schodde, R., 1982. *The Fairy-wrens*. Lansdowne Editions, Melbourne.

• LOGRUNNERS (P. 198)

Boles, W.E., 1988. *The Robins and Flycatchers of Australia*. Angus & Robertson, Sydney.

Keast, A., H.R. Recher, H. Ford & D. Saunders (eds), 1985. *Birds of Eucalypt Forests and Woodlands: Ecology, Conservation and Management*. Surrey Beatty & Sons & RAOU, Melbourne.

• MONARCHS & THEIR ALLIES (P. 199)

Campbell, B.U., & E. Lack (eds), 1985. *A Dictionary of Birds*. T & A.D. Poyser, Calton, UK. pp. 358-359.

• TITS (P. 202)

Barnes, J.A.G., 1975. *The Titmice of the British Isles*. David & Charles, Newton Abbott.

Gaston, A.J., 1973. 'Ecology and Behaviour of the Long-tailed Tit', *Ibis, 115*: 330-51.

Harrap, S., & D. Quinn, 1996. *Tits, Nuthatches and Treecreepers*. Christopher Helm, London.

• NUTHATCHES & TREECREEPERS (P. 204)

Harrap, S., & D. Quinn, 1995. *Tits, Nuthatches & Treecreepers*. Christopher Helm, London.

• HONEYEATERS & THEIR ALLIES (P. 205)

Longmore, N.W., 1991. *Honeyeaters and their Allies of Australia*. Angus & Robertson, Sydney.

Delacour, J., 1944. 'A revision of the family Nectariniidae (sunbirds)', *Zoologica, 29*: 17-38.

Moreau, R.E., 1957. 'Variation in the western Zosteropidae (Aves)', *Bulletin British Museum (Natural History) Zoology, 4*: 311-433.

Pyke, G.H., 1980. 'The foraging behaviour of Australian Honeyeaters: a review and some comparisons with hummingbirds', *Australian Journal of Ecology, 5*: 343-369.

• VIREOS (P. 209)

Bent, A.C., 1950. 'Life Histories of North American Wagtails, Shrikes, Vireos, and their Allies', *US National Museum Bulletin, no. 197* (reprinted by Dover Publications, New York).

• BUNTINGS & TANAGERS (P. 210)

Byers, C., U. Olsson & J. Curson, 1995. *Buntings and Sparrows. A Guide to the Buntings and North American Sparrows*. Pica Press, Sussex.

Isler, M.L., & P.R. Isler, 1987. *The Tanagers: Natural History, Distribution, and Identification*. Smithsonian Institution, Washington D.C.

Skutch, A.F., 1989. *Life of the Tanager*. Comstock, Ithaca.

• WOOD WARBLERS & ICTERIDS (P. 214)

Curson, J., D. Quinn & D. Beadle, 1994. *Warblers of the Americas: an identification guide*. Houghton Mifflin Co., Boston.

Dunn, J., & K. Garrett, 1997. *A Field Guide to Warblers of North America*. Houghton Mifflin Co., Boston.

Morse, D.H., 1989. *American Warblers: an ecological and behavioral perspective*. Harvard University Press, Cambridge, Massachusetts.

Nero, R.W., 1984. *Redwings*. Smithsonian Institution Press, Washington, D.C.

Orians, G.H., 1985. *Blackbirds of the Americas*. University of Washington Press, Seattle.

• FINCHES (P. 216)

Clement, P., A. Harris & J. Davis, 1993. *Finches and Sparrows: An Identification Guide*. Princeton University Press, Princeton.

Goodwin, D., 1982. *Estrildid Finches of the World*. British Museum (Natural History), London.

Restall, R., 1996. *Munias and Mannikins*. Pica Press, Mountfield.

Summers-Smith, D., 1988. *The Sparrows: A Study of the Genus* Passer. T. & A.D. Poyser, Calton.

• STARLINGS & THEIR ALLIES (P. 223)

Amadon, D., 1956. 'Remarks on the starlings, family Sturnidae', *American Museum Novitates, 1247*: 1-41.

Peters, J.L., 1962. *Check-list of birds of the world, vol. 15*. Harvard University Press, Cambridge, Massachusetts.

Vaurie, C., 1949. 'A revision of the bird family Dicruridae', *Bulletin American Museum Natural History, 93*: 199-342.

• NEW ZEALAND WATTLEBIRDS (P. 225)

Greenway, J.C., Jr, 1967. *Extinct and vanishing birds of the world, 2nd edn, revised*. Dover Publications, New York.

• MAGPIE-LARKS & THEIR ALLIES (P. 226)

Heinsohn, R., 1995. 'Raid of the Red-eyed Chicknappers', *Natural History, 104, part 2, Feb. 1995*.

McEvey, A., 1976. 'Osteological notes on Grallinidae, Cracticidae and Artamidae', pp. 150-160 in *Proceedings of the 16th International Ornithological Congress, 1974*. Australian Academy of Science, Canberra.

Rowley, I., 1975. *Bird Life*. Collins, Sydney. 284 pp.

• BOWERBIRDS & BIRDS OF PARADISE (P. 228)

Cooper, W.T., & J.M. Forshaw, 1977. *The Birds of Paradise and Bower Birds*. Collins, Sydney.

Frith, C.B., & B.M. Beehler, 1998. *The Birds of Paradise*. Oxford University Press, Oxford.

Gilliard, E.T., 1969. *Birds of Paradise and Bowerbirds*. Weidenfield & Nicolson, London.

• CROWS & JAYS (P. 232)

Cramp, S., & C.M. Perrin (eds), 1994. *The Birds of the Western Palearctic. Vol. VIII. Crows to Finches*. Oxford University Press, Oxford.

Goodwin, D., 1986. *Crows of the world, 2nd edn*. British Museum (Natural History), London.

Madge, S., & H. Burn, 1994. *Crows and Jays: A Guide to the Crows, Jays and Magpies of the World*. Christopher Helm, London.

Smythies, B.E., 1980. *Birds of Borneo, 3rd edn*. Malayan Nature Society

INDEX

ACKNOWLEDGMENTS

The editors and publishers would like to thank the following people for their assistance and support: Dr Walter Boles, Australian Museum; Graeme Phipps, Curator of Birds, Taronga Zoo, Sydney, Australia; Beverley Barnes; Jenny Mills; Tristan Phillips; and Alison Pressley.